FORENSIC MEDIA

 Sign, Storage, Transmission

A series edited by Jonathan Sterne and Lisa Gitelman

Greg Siegel

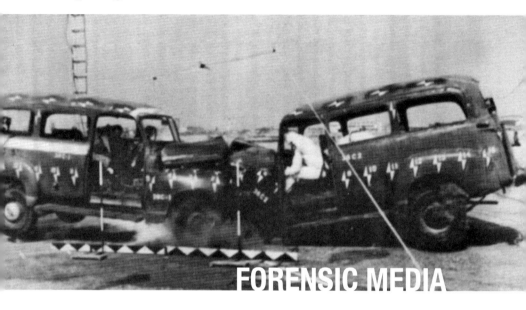

FORENSIC MEDIA

Reconstructing Accidents in Accelerated Modernity

Duke University Press   Durham and London   2014

© 2014 Duke University Press. All rights reserved
Printed in the United States of America on acid-free paper ∞
Designed by Courtney Leigh Baker
Typeset in Helvetica Neue and Whitman by Graphic
Composition, Inc., Bogart, Georgia

Library of Congress Cataloging-in-Publication Data
Siegel, Greg
Forensic media : reconstructing accidents in accelerated
modernity / Greg Siegel.
pages cm—(Sign, storage, transmission)
Includes bibliographical references and index.
ISBN 978-0-8223-5739-1 (cloth : alk. paper)
ISBN 978-0-8223-5753-7 (pbk. : alk. paper)
1. Disasters—Press coverage. 2. Transportation accidents—
Investigation. 3. Forensic sciences. I. Title. II. Series: Sign,
storage, transmission.
PN4874.D57S55 2014
363.12'065—dc23   2014018936

Cover art: Crash-damaged flight recorder. Cover photograph by
and courtesy of Jeffrey Milstein.

To my parents, Nancy and M Barry

The history of human knowledge has so uninterruptedly shown that to collateral, or incidental, or accidental events we are indebted for the most numerous and most valuable discoveries, that it has at length become necessary, in prospective view of improvement, to make not only large, but the largest, allowances for inventions that shall arise by chance, and quite out of the range of ordinary expectation. It is no longer philosophical to base upon what has been a vision of what is to be. *Accident* is admitted as a portion of the substructure. We make chance a matter of absolute calculation.

—EDGAR ALLAN POE, "The Mystery of Marie Rogêt" (1842)

contents

ACKNOWLEDGMENTS ix

INTRODUCTION. Accidents and Forensics 1

one. Engineering Detectives 31

two. Tracings 65

three. Black Boxes 89

four. Tests and Split Seconds 143

EPILOGUE. Retrospective Prophecies 195

NOTES 215

BIBLIOGRAPHY 237

INDEX 251

acknowledgments

The first glimmer of this book came when, during a grad seminar at the University of North Carolina, Chapel Hill, I heard a presentation about the cockpit voice recorder by Mark Robinson, a gifted audio artist. While the particulars of his creation now escape me, I remember being fascinated by the way he deftly integrated actual black-box recordings into an intricate sonic assemblage. The same seminar introduced me to Paul Virilio's work on the accident. These were the seeds.

At the University of North Carolina, I benefited from the wisdom and support of some outstanding mentors and friends. Larry Grossberg, who guided me intellectually and professionally during those green years, taught me most of what I know about cultural studies and the philosophy of communication. He also nurtured my ability to draw clear conceptual distinctions and to engage in rigorous critical analysis. What instances of clarity and rigor are to be found in these pages owe much to his influence. Tyler Curtain, Ken Hillis, Kevin Parker, Della Pollock, and Barbara Herrnstein Smith (at Duke University) each helped me to develop and refine the ideas herein, and I am grateful to them. As for my UNC comrades, Gwen Blue, Steve Collins, Andrew Douglas, Rivka Eisner, Nathan Epley, Mark Hayward, Mark Olson, Phaedra Pezzullo, Bob Rehak, Jonathan Riehl, and Matt Spangler offered encouragement during the formative years. Special thanks and praises to Josh Malitsky, Jules Odendahl-James, and Ted Striphas—three friends whose sage counsel, keen insights, and close fellowship made *Forensic Media* thinkable.

To say that my colleagues in the film and media studies department at the University of California, Santa Barbara, have been exceptionally generous and supportive is to barely scratch the surface. In ways subtle and profound, Edward Branigan, Peter Bloom, Michael Curtin, Anna Everett, Dick Hebdige, Jennifer Holt, Ross Melnick, Lisa Parks, Constance Penley, Bhaskar Sarkar, Cristina Venegas, Janet Walker, and Chuck Wolfe—world-class scholars and wonderful people all—inspired me, and emboldened me, to see this book through. Extra gratitude to Jenny for doing her damnedest to keep me on firm footing (amid the ups) and in good humor (amid the downs). Thanks also to Kathy Murray, department manager extraordinaire, for her always amiable replies to my many pesky questions.

Other luminaries at UCSB provided intellectual stimulation, interdisciplinary opportunities, and professional guidance during the writing of this book. I am indebted to Bishnu Ghosh, Lisa Hajjar, Wolf Kittler, Patrick McCray, David Novak, Rita Raley, and Russell Samolsky for lending a hand or pointing a way. I am indebted, as well, to Meredith Bak and Abby Hinsman for their astute, resourceful, and dedicated research assistance. I thank UCSB's College of Letters and Science, the Hellman Family, and UCSB's Interdisciplinary Humanities Center for the faculty fellowships I received while researching and writing this book.

For their correspondence or other assistance, I am grateful to Karen Beckman, John Brockmann, René Bruckner, James Cahill, Ken Carper, Scott Curtis, Oliver Gaycken, Lisa Gitelman, Dennis Grossi, James Hay, Sarah Lochlann Jain, Akira Lippit, Colin Milburn, David Morton, Sina Najafi, Paul Niquette, Jeremy Packer, Jussi Parikka, John Durham Peters, Henry Petroski, Raymond Puffer, Eric Schatzberg, Jonathan Sterne, Marion Sturkey, Eyal Weizman, Gerald Wilson, Patrick Wright, and Peter van Wyck. I am incredibly lucky that Karen and Jonathan, in particular, took an early and sustained interest in this project. To have two such brilliant and accomplished scholars in my corner—well, it is hard to express in a few words how much their advice, generosity, and endorsement have meant to me. Each deserves more thanks than I have space here.

If I tried the patience of Courtney Berger, my editor at Duke University Press, she was gracious enough not to show it. Her thoughtful comments and steady navigation made this a better book. Thanks also to Erin Hanas at Duke University Press for her superb editorial assistance, to Danielle Szulczewski for shepherding the book through the production process, and to Ken Wissoker for his support.

I am grateful to Ken Goldberg for permission to use his *Dislocation of Intimacy* image, to Jennifer McDaid at Norfolk Southern Corporation for help with the train-wreck photos in chapter 1, and to Erin Rushing at the Smithsonian for help with the *Phonogram* mermaid illustration. I am especially thankful to the photographer Jeffrey Milstein for generously allowing me to use his amazing flight-recorder image for this book's cover—an image that perfectly evokes the medium's danger aura and damaged beauty alike. More of Jeff's remarkable work can be viewed at www.jeffreymilstein.com.

Finally, this book would not have been possible without the moral support of friends and family. Encouragement from Jackie Apodaca, Steve Baltin, Betsy Berman, Julie Bowden, Rick Butler, Andrew Dickler, Debbie Kahler Doles, Jeff Fishbein, Sheila Flaherty, David Greenberg, Rick Habor, Pete Howard, Ned Jennison, Gary Komar, Jon Leaver, Craig Leva, David Marcus, James O'Brien, Tyke O'Brien, Dale Sherman, and my sister Ali Leigh enabled me to keep keepin' on. My deepest gratitude is reserved for my parents, Nancy and Barry. I am beyond fortunate that they have always believed in me and have always been there for me. With love, I dedicate this book to them.

An earlier version of chapter 2 appeared as "Babbage's Apparatus: Toward an Archaeology of the Black Box," *Grey Room* 28 (2007): 30–55. Portions of the introduction and chapter 4 appeared as "The Accident Is Uncontainable/The Accident Must Be Contained: High-Speed Cinematography and the Development of Scientific Crash Testing," *Discourse* 30, no. 3 (2008): 348–72.

introduction

ACCIDENTS AND FORENSICS

The will to mastery becomes all the more urgent the more technology threatens to slip from human control. —MARTIN HEIDEGGER, "The Question Concerning Technology"

In the disaster milieu, technological civilization is on trial as it attempts to heal the systemic breach and restore itself through a figural elimination of all risk. Various cultural interpretations, analyses, and judgments attempt reconstruction of a safe world without slippage, broadly defined. —ANN LARABEE, *Decade of Disaster*

Whatever Can Go Wrong

From 1947 to 1951, a series of groundbreaking experiments was conducted at Edwards Air Force Base in California's Mojave Desert.[1] The objective of the military research project was to study the limits of human tolerance to rapid deceleration as well as the strength of airplane seats and harnesses under simulated crash conditions. A rail-mounted, rocket-boosted sled with a high-powered hydromechanical braking system was employed to this end. The U.S. Air Force officer and flight surgeon John Paul Stapp directed the audacious experiments, and he himself sometimes served as a test subject.[2]

After completing one such experiment in 1949, researchers discovered to their dismay that the electrical sensors affixed to Stapp's safety harness had failed. Instead of a positive numerical value, each sensor yielded a zero reading. Perplexed, the team of experimenters sought to ascertain the cause of the failure. Why had the instruments malfunctioned? What had gone wrong? It was soon revealed that the sensors had been installed improperly—backward, in

John Paul Stapp aboard rocket sled, circa 1947. Note the accelerometers on his knee and chest and in his mouth. Courtesy of Edwards Air Force Base History Office.

fact. On learning of the mishap, Captain Edward A. Murphy Jr., one of the test technicians, quipped in frustration, "Whatever can go wrong, will go wrong." Thus was born Murphy's Law.

While not as widely known as the adage itself, the Murphy's Law origin story has, over the decades, become something of a minor legend.[3] As with all legends, the story circulates in many "true" versions. Was Stapp the one who rode the rocket sled, or was the test subject actually a chimpanzee? How many sensors were involved—four? Six? Sixteen? Were they affixed to Stapp's harness or to his body? Were they mounted incorrectly, or were they defective from the outset? Was Murphy at fault, or was another technician to blame? Did Murphy coin the expression on the spot, or did Stapp do so days later at a press conference? And how, exactly, was the expression worded? Each of the following phrasings has been claimed as accurate: "If there's any way they [the team of technicians] can do it wrong, they will"; "If anything can go wrong, it will"; "If anything can go wrong, he [Murphy] will do it"; "If it can happen, it will";[4] "If that guy [Murphy's assistant] has any way of making a mistake, he will"; "If there's more than one way to do a job, and one of those ways will result in disaster, then somebody will do it that way"; "Whatever can go wrong, will go wrong."

Which, if any, version of the origin story is historically true and which, if any, variant of the adage is historically accurate are less important for our purposes here than the cultural perceptions they collectively articulate. Murphy's Law makes a popular truism out of a pessimistic fatalism, extracts a nugget of folk wisdom from a philosophy of despair. It presumes the failure of every endeavor and, in a sense, predicts the worst of all possible worlds (and does so with a hint of perverse delight). It insists that, in the future, mistakes, misfortunes, and other inauspicious outcomes are not only probable but, indeed, unavoidable. It says that the proverbial best-laid plans *always* go astray.

Perhaps more intriguing, though, is the singular way in which Murphy's Law evinces a certain attitude toward modern technology. Interpreted in the light of its origin story, with its faulty sensors and frustrated scientists, the enduring maxim expresses not so much a sweeping cosmological pessimism as a narrow technological one. "Whatever can go wrong, will go wrong" effectively translates as "However technology can fail, it will fail." To be sure, the mythology surrounding Murphy's Law reads as a contemporary parable of technological excess and accident, a cautionary tale about the irreducible complexity of human-machine relations and the unruly contingencies that vex—and seem to hex—them.

Yet even this reading does not quite reach the story's deeper meaning, which is attained only when a grand irony is grasped. The irony is this: Stapp's rapid-deceleration research was undertaken with the goal of minimizing the injurious effects of high-speed accidents. Because the experiments were complicated and fraught with difficulty and danger, extraordinary steps were taken to ensure the smoothness of operations and the safety of test subjects. Still, the accident proved irrepressible. The moral of the Murphy's Law myth? Its devastating subtext? The accident thwarts even the most technologically advanced attempts to tame it. Its demons possess the power to disturb even the scene of their own exorcism.

A Mystery That Riveted the World

In the middle of the night on 1 June 2009, Air France Flight 447, en route to Paris from Rio de Janeiro, vanished over the Atlantic Ocean, somewhere between South America and Africa. The pilots were last heard from at

1:35 A.M., when they made routine radio contact with Brazil Air Traffic Control. The plane, an Airbus A330 carrying 228 persons, was last spotted on Brazilian radar screens thirteen minutes later.[5] And then, nothing—no further sight or sound. Why did the two-hundred-ton jetliner suddenly disappear? What caused it to eventually crash into the ocean, killing everyone aboard? What really happened? The *New York Times* called the incident "a mystery that riveted the world."[6]

Over the next few weeks, French and Brazilian search teams, dispatched to scour the Atlantic for wreckage, would recover more than six hundred fragments of floating debris and the bodies or body parts of some fifty individuals.[7] Authorities from France's Bureau of Investigations and Analyses hoped that such technological and human remains would, after thorough forensic examination, help them piece together the tragic events leading up to the catastrophe. But what they needed most of all—and still did not possess or even know the location of—were the airplane's flight-data recorder and cockpit voice recorder, its two "black boxes." Absent the evidence contained in these devices, it would be all but impossible to make meaningful sense of the accident. As the *New York Times* reported, "Without the secrets locked in the recorder[s'] hard drive[s]—conversations of the crew in the flight's final moments, crucial data on the plane's altitude, airspeed and heading—the answers to what exactly happened may forever be entombed in the sea, along with the majority of the crash victims."[8]

Nearly two years later, after several failed searches, a good deal more wreckage—including engines, landing gear, a section of fuselage, and 104 more bodies—was discovered, not, this time, bobbing on the surface, but rather resting heavy on the ocean bottom.[9] The astonishing find was accomplished through technical means and feats that were themselves nothing short of astonishing. In early April 2011, a REMUS 6000, a state-of-the-art unmanned reconnaissance submarine equipped with side-scan sonar transducers and a high-resolution digital camera synchronized with strobe light, delivered to scientists involved in the recovery effort thousands of black-and-white images of the seafloor debris field.[10] While the discovery represented a major breakthrough, air-crash investigators were nonetheless disappointed that the black boxes—"the salvage operation's first priority"—remained missing.[11]

Less than a month later, the drone submarine at last found the flight-data recorder and the cockpit voice recorder, and by mid-May the information stored therein had been "successfully downloaded . . . and transferred onto a secure computer server." Of this practically miraculous retrieval, a French transportation minister proclaimed, "This proves we were right to devote such an effort

REMUS 6000 ready to launch. Photograph by and courtesy of Michael Dessner.

to shed light on this accident."[12] Aviation-safety experts vowed to conduct "a detailed analysis of the black box recordings in order to assemble a fuller narrative of what happened."[13] The colossal two-year investigation had already cost Air France, Airbus, and the French government collectively $30 million. More evidence, new knowledge, "a fuller narrative," a properly scientific explanation: these were the dividends the flight recorders promised to return on this considerable investment of public and private resources.

On 5 July 2012, the French Bureau of Investigations and Analyses released its long-awaited final report on Air France Flight 447. Its official conclusions were derived in large measure from the messages (technical, linguistic, sonic) encoded and preserved in the airplane's black boxes. As the report states, "The investigation into the accident to AF 447 confirms the importance of data from the flight recorders in order to establish the circumstances and causes of an accident and to propose safety measures that are substantiated by the facts."[14]

Accidents and Forensics 5

Let us notice the dual assertion here. The signs, words, and sounds electronically etched in the recorders' memory units have the power to illuminate the accident's causes in the present, *and* they hold the potential to help prevent accidents in the future. Such information can be used to solve the "world-riveting mystery" now *and* to save lives later. Black-box knowledge enables the retrospective revelation of empirical facts *and* the prospective protection of vulnerable bodies (and breakable machines).

Forensic Mediation and Modernity

Taken together, John Paul Stapp's rocket-sled tests (including the botched one that allegedly gave birth to Murphy's Law) and the more recent episodes involving Air France Flight 447's black boxes (the extended search, the expensive recovery, the expert decipherment) point to a world in which modern technology by turns blesses and curses, ruins and restores, enlightens and bewilders, "slips from human control" and supplies the means to secure itself anew.

They also illustrate quite vividly the extraordinary lengths to which technologically modern societies will go to learn about—and to learn from—accidents and failures: the costs they will bear to plumb their murky depths, the complexities they will negotiate to divine their secret truths, the risks they will accept to receive their apparently oracular wisdom. Stapp's crash-injury research at Edwards (and, later, at Holloman Air Force Base in New Mexico) used electrical sensors and cables, radiotelemetry equipment, high-speed photographic and cinematographic cameras, and other sophisticated instrumentation to gather and analyze data. The Air France Flight 447 accident investigators relied heavily on the digital and acoustic evidence housed in, and painstakingly extracted from, the plane's flight-data recorder and cockpit voice recorder—recorders whose deep-sea salvage itself marked a major technoscientific achievement. Today, such elaborate crash-experimentation programs and crash-investigation procedures are well established and widespread throughout the industrialized world, routinely expected and ritually obligatory aspects of our modern "disaster milieu." And here we come to the crux of the matter as far as the present study is concerned. On the one hand, simulated accidents are meticulously orchestrated in order to technically register and measure them as they happen. On the other hand, real accidents are methodically investigated in order to logically "reconstruct" (this the term of art) them after the fact. In both cases, high-speed mishaps become objects of scientific and governmental inquiry—and of popular imagination—through devices and protocols of *forensic mediation*.[15]

Broadly, *Forensic Media* looks at the interrelation of accidents, forensics, and media in the culture of modernity, where "modernity" can be provisionally understood, following Matei Calinescu, to denote "a stage in the history of Western civilization—a product of scientific and technological progress, of the industrial revolution, of the sweeping economic and social changes brought about by capitalism."[16] More to the point, this book is about how, since the nineteenth century, media technologies have informed and facilitated an ongoing project to deal with the problem of technological accidents, particularly high-speed crashes and catastrophes. Forensic-scientific in nature and method, this ongoing project, like the problem it enunciates, has multiple intersecting dimensions: cultural and institutional, practical and epistemological, material and ideological. My overarching contention in these pages is that accidents, forensics, and media are mutually implicated in the origins and evolution of a dominant tendency in modern technological thought, discourse, and practice. This tendency treats forensic knowledge of accident causation as the key to the accident's solution, the rational answer to its constitutive riddle. It also treats such knowledge as a source of future technical improvement and, by further (and somewhat fantastic) extension, of future sociotechnological advancement, of progress on a civilizational scale. *Forensic Media* examines this peculiar complex of scientific attitude and cultural mythos by considering the ways and contexts in which graphic, photographic, electronic, and digital media have been adapted and deployed to informationalize, anatomize, and narrativize accidents of accelerated mobility. Throughout, I show how such devices have been pressed into service to forensically work on, work out, and work through such disasters: to scientifically detect and inspect them, to epistemically manage and discursively control them. In offering a new account of the historical links and cultural relays between accidents and forensics, I ultimately aim to tell a new story about the corresponding connections between media, technology, and modernity.

Forensic Media's case studies and analyses are organized around a set of critical questions: how have devices of recording, representation, and reproduction been employed to scientifically analyze and explain high-speed mishaps? What was the historical impetus for so employing them? How do the imagination and the practice of reconstructing accidents through aural, visual, and audiovisual media enact a distinctly forensic rationality and epistemology? How do such cultural imaginings and institutional practices embody a larger forensic project and ideology? How do certain forensic media—namely, Charles Babbage's "self-registering apparatus" (chapter 2), flight-data recorders and cockpit voice recorders (chapter 3), crash-test cinematography (chapter 4), and accident-reconstruction technologies (epilogue)—rearticulate other, older or concurrent media forms, discourses, and applications? How do they express a particular range of psychosocial desires and anxieties, hopes and fears? How do they function as instruments of scientific and governmental investigation and, at the same time, as vehicles of popular communication? How do they inculcate or reinforce a specific ethos of safety and protection? Of precaution and accident prevention? What role have they played in shaping modern perceptions of accidents and failures, especially of their causes and circumstances? Finally, and most speculatively, how have they contributed to a latter-day revision of the notion of progress?

I explore these questions in the chapters that follow. First, however, I here outline and begin to explicate forensic media's philosophical conditions of possibility, along with their historical and discursive contexts of emergence.

The Problem of the Accidental

"Philosophical reflection has no other object than to get rid of what is accidental."[17] If he overstates the case, G. W. F. Hegel is nevertheless not fundamentally wrong: the accidental and its corollaries—fortuity, contingency, indeterminacy—have troubled Western thought and discourse since Greek antiquity.

The trouble began with Aristotle, who banished chance occurrences, or "accidents," from his orderly metaphysical system because of their absolute particularity and practically infinite variability. Centuries later, the Roman philosopher Boethius tried to harmonize Aristotle's position on accidents with Christian teachings on divine providence, which, as Michael Witmore notes, "helped subsequent theologians to embrace the contradictory notion that God foresaw and controlled accidents without taking an active role in bringing them about."[18] Jean Calvin and his English interpreters in the early modern period, by contrast, strenuously rejected the suggestion that the Christian deity was not the immediate author of ostensibly fortuitous events. According to Calvin's doctrine of "special providence," each and every mundane occurrence, "no matter how unexpected or inconsequential," manifests the efforts of a divine micromanager.[19] All things that happen on earth, as in the heavens, are directly attributable to a "disposing" God; they constitute both the purposeful results of His action and (for those with eyes to see and ears to hear) the wonderful signs of His intervention. Hence, on this conception, there are no, and can be no, wholly contingent phenomena in the universe, no true luckiness or unluckiness, no accidents per se. Indeed, the very category of the accidental is here dismissed as a metaphysical or "theological absurdity."[20]

Philosophical system builders and Christian providentialists are not the only ones to categorically refuse the accidental, or so it has been authoritatively claimed. In his treatise *The Child's Conception of Physical Causality* (1927), the developmental psychologist Jean Piaget argued that young children, like Aristotle (the comparison is Piaget's), feel "a very definite repugnance" for the idea of chance:

> To our eyes, nature is simply the totality of necessary sequences and of their interferences, which interferences characterise what we call chance. For the child, on the contrary, the world is a realm of ends, and the necessity of laws is moral rather than physical. . . . The very way in which the child frames his questions before the age of 7–8 is evidence of an implicit belief in a world from which all chance is proscribed. . . . Sun and moon, cloud and wind are always supposed, whatever they may do, to be acting with some intention or other. Their will may be capricious, but there is no element of chance in the transaction.[21]

Five years earlier, the philosopher and armchair anthropologist Lucien Lévy-Bruhl described the "primitive mentality" in much the same way:

> From disease and death to mere accidents is an almost imperceptible transition. . . . Primitives, as a rule, do not perceive any difference between a death which is the result of old age or of disease, and a violent death. . . . Therefore every death is an accidental one, even death from illness. Or to put it more precisely, no death is, since to the primitive mind nothing ever happens by accident, properly speaking. What appears accidental to us Europeans is, in reality, always the manifestation of a mystic power which makes itself felt in this way by the individual or by the social group.[22]

Though he disputed some of Lévy-Bruhl's theoretical conclusions, the colonial anthropologist E. E. Evans-Pritchard, in his classic study of the Azande of central Africa, published in 1937, likewise distinguished between "civilized" and "uncivilized" peoples on the basis of their differing interpretations of the causes of misfortune.[23] Whereas the former tend to be interested in *how* accidents happen (a technical or scientific question), the latter, according to Evans-Pritchard, are inclined to explain *why* they happen to whom they do (a social and moral question). For the Azande, such explanations typically involved the concept of witchcraft. "Witchcraft is a causative factor in the production of harmful phenomena in particular places, at particular times, and in relation to particular persons. It is not a necessary link in a sequence of events but something external to them that participates in them and gives them a peculiar value."[24] Evans-Pritchard offers the example of an old granary that collapses due to termite infestation, injuring the people working and conversing beneath its roof:

> We say that the granary collapsed because its supports were eaten away by termites. That is the cause that explains the collapse of the granary. We also say that people were sitting under it at the time because it was in the heat of the day and they thought that it would be a comfortable place to talk and work. This is the cause of people being under the granary at the time it collapsed. To our minds the only relationship between these two independently caused facts is their coincidence in time and space. We have no explanation of why the two chains of causation intersected at a certain time and in a certain place, for there is no interdependence between them.
>
> Zande philosophy can supply the missing link. The Zande knows that the supports were undermined by termites and that people were sitting beneath the granary in order to escape the heat and glare of the sun. But he knows besides why these two events occurred at a precisely similar moment in time and space. It was due to the action of witchcraft. If there had been no witchcraft people would have been sitting under the granary and it would not have fallen on them, or it would have collapsed but the people would not have been sheltering under it at the time. Witchcraft explains the coincidence of these two happenings.[25]

In these scholarly accounts from the early twentieth century, child and "primitive" alike inhabit a maximally enchanted world—a world naturally replete with moral significance and supernaturally devoid of mere coincidence. Just as the child's proscription of chance is simultaneously an index of his immaturity and a function of his pre-rationality, so the primitive's nonconception of the accidental is a token of both his unscientificity and his uncivility—indeed, is regarded as the very mark and measure of his distance from a normative Western modernity. For Lévy-Bruhl and Evans-Pritchard, the ability to apprehend the ordinary workings of contingency, to recognize in certain occurrences an irreducible *accidentality*—to acknowledge, in short, the fact that some things "just happen"—is precisely an achievement of reason and scientific enlightenment. Belief in the element of chance, acceptance of mere coincidence, serves as a standard of intellectual sophistication and an expression of cultural superiority. Primitive peoples, what with their "superstitious" cosmologies, imagine that magical or animistic forces are necessarily implicated in the realization of nearly every unfortunate event. Modern populations ("us Europeans"), by contrast, having renounced the myth of supernatural influence, are supposed to understand that much of what happens in the world does so without intention or motivation and, therefore, contains no intrinsic meaning or purpose. Epistemological modernity makes accidents possible as such.

Judith Green has located the accidental's modern provenance in the fissure between two articles of rationalist faith, in the slippage between two prevailing scientisms.[26] The first of these is the belief in the laws of physical causation vis-à-vis the production of events; the second, the belief in the laws of statistical probability vis-à-vis the distribution of those events. On the one hand, modern reason avows that every misfortune, like every other occurrence (willed or not, foreseen or not), is the effect of a concrete, determinate cause or chain of causes. On the other hand, it knows that any particular misfortune, any specific instance of "bad luck," is (or was, before it came to pass) always only a mathematical possibility, not a preordained certainty, not "destiny." After the Enlightenment, the philosophical and discursive adhesive that holds these two propositions together, that joins causal determinism to statistical probabilism, physical necessity to phenomenal contingency, takes the name "accident." "An accident appears in the gaps left by a rationalist cosmology, at the limit of deterministic laws, but where superstition no longer has a legitimate part to play," writes Green in *Risk and Misfortune*. "These gaps in rationalist explanations emerge between what is known for sure (that is, that which is subject to deterministic laws, such as those describing the motions of planets around the sun, or gravity on the earth) and that which is known statistically (that is, that which is subject to the laws of probability, such as the chance of reaching a certain age or of dying of a certain disease)."[27] In a similar vein, Octavio Paz contends that "the Accident has become a paradox of necessity: it possesses the fatality of necessity and at the same time the indetermination of freedom."[28]

At once excessive to the system and, paradoxically, like the Derridian "supplement," an excess the system requires for self-coherence and self-completion, the "residual category" of the accidental is crucial to modern thought and discourse because it functions to stabilize the fault line between deterministic and probabilistic logics, without resorting to interpretations rooted in either providential or "primitive" worldviews.[29] "To classify," declares Zygmunt Bauman in *Modernity and Ambivalence*, "is to give the world a *structure*: to manipulate its probabilities; to make some events more likely than some others; to behave as if events were not random, or to limit or eliminate randomness of events."[30]

The accidental is ideologically essential to modernity, then, because it allows the "civilized mentality" to rationally classify those irruptions of randomness it cannot really explain.

The Problematization of the Accident

So the accidental proposes to solve something of a doctrinal problem for modern cosmology. Yet modernity by no means escapes the problem of the accident—not materially, not symbolically, not affectively. On the contrary, since the dawn of the Industrial Revolution—since, that is, the accident entered what Thomas Carlyle, in 1829, dubbed "the Mechanical Age," or "the Age of Machinery"—the problem has become virtually omnipresent, even as it has acquired radically novel characteristics and significances.[31] To be sure, through its encounter with modern technology, the ancient problem of the accident has been fundamentally transformed, taking on fresh urgency and also far greater complexity. At the same time, it has assumed a strange and terrible aspect, its brutal reality reflected in every industrial mishap, its violent intensity exhibited in every high-speed collision, its frightful enormity demonstrated in every engineering disaster. If modernity needs the category of the accidental, it nevertheless cannot stand—indeed, it dreads and is desperate to avoid—technological accidents.

The accident became technologically modern in the nineteenth century. During this tumultuous era, the headlong development, capitalization, and wide-scale implementation of new sources of energy and motive power—mainly steam and electricity—injected into everyday life a bewildering multitude of new hazards and vulnerabilities, both physical and psychoperceptual. In mines, mills, and factories; aboard steamboats, railroad trains, and trolley cars (and, before long, motorcars); in shops, offices, and theaters; on city streets, docks, and bridges; in public buildings and even private residences—nearly everywhere, it seemed, human bodies were exposed to unprecedented threats and dangers, while human sense organs were bombarded by unfamiliar shocks and uncanny disturbances, all resulting from the processes and products of industrialization, mechanization, and electrification. A colorful passage from *The Education of Henry Adams*, first printed in 1907, captures the state of affairs: "Every day Nature violently revolted, causing so-called accidents with enormous destruction of property and life, while plainly laughing at man, who helplessly groaned and shrieked and shuddered, but never for a single instant could stop. The railways alone approached the carnage of war; automobiles and firearms ravaged society, until an earthquake became almost a nervous relaxation.

An immense volume of force had detached itself from the unknown universe of energy, while still vaster reservoirs, supposed to be infinite, steadily revealed themselves."[32]

Adams's literary description of modern technology's "accidental" inflictions and innervations, besides resonating with certain of the urban sociological writings of Georg Simmel, Siegfried Kracauer, and Walter Benjamin, paralleled the sensational depictions of metropolitan modernity in the fin-de-siècle press. As Ben Singer observes, "The sense of a radically altered public space, one defined by chance, peril, and shocking impressions rather than by any traditional conception of continuity and self-controlled destiny," pervaded the illustrated newspapers—and, with them, the popular imagination—of the day. "Unnatural death . . . had been a source of fear in premodern times as well . . . , but the violence, suddenness, [and] randomness . . . of accidental death in the metropolis appear to have intensified and focalized this fear."[33]

The novel causes and fearsome circumstances of accidental injury and death in the nineteenth century prompted a broad range of social, cultural, technical, and institutional responses. In general, these responses were predicated on distinctly modern logics and discourses of risk and safety. Hence the ascendance of a certain dual practico-ideological imperative: minimize/reduce the former, maximize/produce the latter.

Both Ulrich Beck and Anthony Giddens have argued that modernity invests the concept of risk with a unique meaning and with a special epistemological function. In Beck's now-classic formulation, "risk" names "a systematic way of dealing with hazards and insecurities induced and introduced by modernization itself. Risks, as opposed to older dangers, are consequences which relate to the threatening force of modernization and to its globalization of doubt."[34] In contrast to premodern societies, which ascribed calamities and other misfortunes to the gods or demons, to the fates or Fortuna, modern industrialized societies, according to Beck and Giddens, take pains to turn disasters and other unfortunate contingencies into rationally calculable risks. "A world structured mainly by humanly created risks has little place for divine influences, or indeed for the magical propitiation of cosmic forces or spirits," writes Giddens in *The Consequences of Modernity*, sounding very much like Lévy-Bruhl and Evans-Pritchard (sans their colonialist ethnocentrism).[35] "It is central to modernity that risks can in principle be assessed in terms of generalisable knowledge about potential dangers—an outlook in which notions of *fortuna* mostly survive as marginal forms of superstition."[36]

François Ewald, for his part, traces the concept back to the realm of late medieval maritime insurance, wherein "risk designated the possibility of an objective danger, an act of God, a force majeure, a tempest or other peril of the sea that could not be imputed to wrongful conduct."[37] At the start of the nineteenth century, however, "the notion of risk underwent an extraordinary extension: risk was now no longer exclusively in nature. It was also in human beings, in their conduct, in their liberty, in the relations between them, in the fact of their association, in society"—and, we might add, in their relations with technology, in their dealings with modernity's numerous devices and contraptions, machines and modes of conveyance.[38] Ewald continues: "This extension was due in part to the singular appearance of the problem of the accident."[39]

The problem of the accident, indeed. Or, better yet, the *problematization* of the accident. For Michel Foucault, "Problematization doesn't mean representation of a preexisting object, nor the creation by discourse of an object that doesn't exist. It is the totality of discursive or non-discursive practices that introduces something into the play of true and false and constitutes it as an object for thought (whether in the form of moral reflection, scientific knowledge, political analysis, etc.)."[40] The accident was problematized anew in the nineteenth century. To claim this is not simply to claim that the accident appeared or was experienced as a new problem. It is also, and especially, to claim, following Foucault, that the accident—specifically the *technological* accident—was captured by a particular "regime of truth" and, in the same gesture of thought and power, was constructed as a particular object of scientific knowledge and discursive control.[41] It is to maintain that the accident was strategically rendered as—actively made into—a phenomenon susceptible to empirical investigation, methodical dissection, rational explanation, and institutional administration. It is to contend that the question of the accident's nature and operations, of its causative factors and conditions of possibility, became socially enunciated and legitimated, was formally introduced "into the play of true and false."

The emergence and codification of the actuarial sciences provides an illustrative case in point. Building on the seminal work of Ian Hacking, Roger Cooter and Bill Luckin, editors of the volume *Accidents in History*, remark that "the rise of statistics in the mid-nineteenth century secularized accidents as expected events."[42] Governments and insurance companies alike endeavored to stabilize and normalize accidents through processes of "exquisite actuarial reduction," thereby transforming "the once unknowable workings of chance into calculations that reinforce[d] the predictability of everyday economic and technological life."[43]

Over and above the abstruse reckonings of actuarial rationality, the accident's modern problematization was, and continues to be, manifest in countless safety rhetorics and rituals, artifacts and institutions. No doubt, the nineteenth and twentieth centuries witnessed the birth and diffusion of a new precautionary ethos and consciousness, embodied in all manner of safety movements and campaigns, safety laws and regulations, safety systems and procedures, safety tools and equipment, safety organizations and cultures, safety experts and professionals. In so-called risk societies, where technological accidents "just happen" and, moreover, are forever "waiting to happen," where (per Singer) they materialize violently, suddenly, and randomly, safety is simultaneously a worthy objective and an everyday obsession, a perfectly reasonable aim and a patently anxious fixation.

"Safety is often considered to be a buffer between technology run rampant and the sanctity of human life," notes Jeremy Packer, "a zone that protects humanity from the modernization machine, the greed of corporations, speed, the crash, the insensitivity of the bureaucratic monster, and breakdowns in the moral order."[44] Since the onset of industrialization, modern subjects and populations have been told time and again that the problem of the accident (the crash, the breakdown, the monster in the machine) has its solution in the provision of safety, in the expansion of that buffer, that zone, that protective space, literal or metaphorical, separating sacred life from risky technology, cushioning the one against the other.

Over the past two centuries, many and various technical "safety devices" have been employed in many and various contexts. Rem Koolhaas, in "'Life in the Metropolis' or 'The Culture of Congestion,'" recounts a legendary story about one exemplary such device:

> In 1853, at Manhattan's first World's Fair, the invention that would, more than any other, become the "sign" of the Metropolitan Condition, was introduced to the public in a singularly theatrical format.
>
> Elisha Otis, the inventor of the elevator, mounts a platform. The platform ascends. When it has reached its highest level, an assistant presents Otis with a dagger on a velvet cushion. The inventor takes the knife and attacks what appears the crucial component of his invention: the cable that has hoisted the platform upward and that now prevents its fall. Otis cuts the cable; nothing happens to platform or inventor.
>
> Invisible safety-catches prevent the platform from rejoining the surface of the earth. They represent the essence of Otis's invention: the ability to prevent the elevator from crashing.
>
> Like the elevator, each technical invention is pregnant with a double image: the spectre of its possible failure. The way to avert that phantom disaster is as important as the original invention itself.[45]

The forensic media I examine in this book may rightly be regarded as safety devices in the Koolhaasian sense. For they, as much as the "technical inventions" toward which they are directed or with which they are integrated, are inevitably haunted by the possibility of failure, shadowed and unsettled by the specter of disaster. Yet forensic media are safety devices of a quite peculiar sort—peculiar in terms of both the kind of safety they promise to deliver and the kind of device they purport to be.

Unknown Causes

To problematize the accident is, first of all, to pose the question of the accident's *cause*. In technologically modern societies, questions of accident causation are investigated *forensically*. They are investigated, that is to say, through the scientific detection and inspection of material traces, the formal decoding of indexical signs, the rigorous analysis of physical evidence. Queries are put to the remains of the catastrophe (including its mediatized remains: its recorded survivals, its "living" reproductions) as though they were a criminal suspect under police interrogation or, still more appositely, a corpse under the disciplinary gaze of an autopsist or pathological anatomist. What really happened? What went wrong? What were the precipitating factors? What were the occasioning circumstances? What was the exact sequence of events, the relevant chain of causes? What, in short, is the *story* of the accident?

The need or desire to search out and understand the causal origins of accidents and unfortunate events—and to quell the panic or terror such occurrences are liable to engender—is, according to both Thomas Hobbes and David Hume, not only a profoundly human but also a primordially religious impulse. "It is peculiar to the nature of man to be inquisitive into the causes of the events they see, some more, some less, but all men so much as to be curious in the search of the causes of their own good and evil fortune," Hobbes proclaims.[46] "The perpetual fear, always accompanying mankind in the ignorance of causes (as it were in the dark), must needs have for object something. And therefore, when there is nothing to be seen, there is nothing to accuse, either of their good or evil fortune, but some *power* or agent *invisible*; in which sense, perhaps, it was that some of the old poets said that the gods were at first created by human fear."[47]

A hundred-odd years after Hobbes's *Leviathan*, Hume, in his *Natural History of Religion*, argued along the same lines, with characteristic eloquence: "We are placed in this world, as in a great theatre, where the true springs and causes of every event are entirely concealed from us; nor have we either sufficient wisdom to foresee, or power to prevent those ills, with which we are continually threatened. We hang in perpetual suspence [sic] between life and death, health and sickness, plenty and want; which are distributed amongst the human species by secret and unknown causes, whose operation is oft unexpected, and always unaccountable. These *unknown causes*, then, become the constant object of our hope and fear."[48] And then, a few paragraphs on: "Every disastrous accident alarms us, and sets us on enquires concerning the principles whence it arose: Apprehensions spring up with regard to futurity: And the mind, sunk into diffidence, terror, and melancholy, has recourse to every method of appeasing those secret intelligent powers, on whom our fortune is supposed entirely to depend."[49]

When bad things happen unexpectedly or unaccountably, these modern philosophers insist, we are anxious to know why. Disaster strikes, and we instinctively strive to dissipate the enveloping epistemic darkness. Nothing, it seems, is so intolerable, so upsetting or threatening, as an unknown cause. Gripped by fear, plagued by ignorance, impelled by curiosity, we go in quest of answers and reassurance. Defensively, we are inclined to investigate the accident's mysterious genesis, its obscure "springs and causes"; desperately, we try to propitiate the invisible superintending intelligence that allegedly dispenses fortune and misfortune, allocates good and evil, weal and woe. "Cause," says Roland Barthes, is "an ironically ambiguous word, referring as it does to a faith and a determinism, as if they were the same thing."[50] Hobbes and Hume alike postulate that the masses of humankind, ever insecure about their uncertain future, have, over the millennia, tended to put their faith in God or the gods, hoping and praying, trusting and believing that every worldly occurrence, favorable or not, has a purposeful cause, and every cause a higher reason.

In the religious imagination, then, the mundane *cause of* any given accident or unfortunate event is identical to the transmundane *reason for* it. The will of God or of the gods is construed as mover, motion, and motive, all at once. "They that believe God dos [sic] not foresee Accidents, because Nothing can be known that is not, and Accidents have no being, untill [sic] they are in Act, are very much mistaken," declares the seventeenth-century man of letters Samuel Butler. "For Accident is but a Terme [sic] invented to relieve Ignorance of Causes, as Physitians [sic] use to call the strange operations of Plants, and Minerals [sic] Occult Qualities, not that they are without their Causes, but that their Causes are unknown. And indeed there is not any thing in Nature, or event, that ha's [sic] not a Pedegree [sic] of Causes, which though obscure to us, cannot be so to God, who is the first Cause of all things."[51]

Modern reason and technology conspire to complicate this theistic picture.

Modern reason secularizes the accident, disenchants it, casts out its primeval ghosts and goblins, while modern technology effectively humanizes it, brands it an anthropogenic failure rather than a "natural disaster" or an "act of God" (those perennial alibis of extraordinary ill chance). If, in modernity, the accident no longer happens "for a reason"—that is, for a *mystic* or *divine* reason—it nevertheless retains a real causal history, a nexus of prior causes governed by the accepted laws of physics. And, crucially, this nexus, this profane "pedigree," is now all there is—all that exists notionally to explain the accident's existence, to comprehend its having come to pass. To be sure, in a disenchanted universe, the accident's reason for being is entirely reducible or equivalent to—is nothing more than, nothing other than—its physical cause of being. Or, to turn the statement around, the accident's non-metaphysical cause is precisely its self-sufficient reason. The question "Why did the accident happen?" is swallowed whole by the question "How did the accident happen?" Witchcraft is devoured by termites.

Much of what Hume said in 1757 about the human experience of "secret and unknown causes" still holds true. "Disastrous accidents" still alarm us, and they still make us apprehensive about the future. We still inquire into their original causal principles, and, in a sense, we still endeavor to appease those "intelligent powers" that supposedly determine our fortune. The difference nowadays, however, is that, the persistence of religious sentiment notwithstanding, science and engineering have developed their own modes of inquiring into the unknown causes of disastrous accidents. What is more, they have developed their own methods of appeasing the gods (the "gods" of technology, the "intelligence" of machinery) as well as their own means of safeguarding the future. Together, these specialized modes, methods, and means can be designated by the term *forensic*.

Dreaming Accident Forensics

The animating impulse of accident forensics, which seeks to positively identify the causes of technological accidents, and that of criminal forensics, which seeks to positively identify the perpetrators of crimes, emerged during same historical period. More than merely historically coincident, in fact, these twin articulations of forensic logic—one applied to mishaps, the other to misdeeds—developed as part of the same constellation of expert knowledges, discourses, and techniques, the same "scientifico-legal complex," to appropriate a term from Foucault.[52] Nineteenth-century modernity not only spawned a new scientifico-legal discipline (criminalistics) in reciprocal association with a new literary fictional genre and protagonist (the detective story, the detective-hero); it also sowed the seeds of that cognate discipline that would, in the twentieth century, come to be called "forensic engineering."

Railroad Accidents: Their Causes and the Means of Preventing Them, written by the French civil engineer Emile With and translated into English in 1856, was probably the first nontechnical treatise to take an incipiently forensic approach to the problem of mechanized-transportation accidents. "For some months, railroad accidents have become so common, that it seems as though science was altogether powerless in preventing them," With asserts. "In truth, they are always occasioned by the imprudence of passengers, the want of foresight in those employed on the road, or by a concurrence of fatal, but very natural circumstances."[53] Proffering rationalist explanations of phenomena ranging from "Explosion of Locomotive Boilers" to "Carelessness of Those in Charge of Engines," from "Defects in the Rolling Stock" to "Want of Communication Between the Conductor and Engine Driver," *Railroad Accidents* concludes that every train wreck can be attributed either to operator error or to a "fatal" (but not fated) conjunction of "very natural" (as against supernatural) forces and conditions. Furthermore, because railroad accidents happen for these, and only these, sublunary reasons, because their causes are traceable often to human behavior and always to brute physics, science is by no means "powerless" to intervene and improve the situation. On the contrary, With assures his readers, by discovering and elucidating the root causes of catastrophes, science demonstrates that "it is never impossible to prevent" the occurrence of similar misfortunes in the future.[54] "In all countries, [railroads] are subjects of study with serious minds, and they are every day advancing more nearly to perfection."[55]

In his preface, With's translator, G. Forrester Barstow, also a civil engineer, protests (as if echoing Samuel Butler in a secular register) that railroad accidents do not truly deserve their common appellation:

> Accidents are defined by Webster to be, "events which proceed from unknown causes, or unusual effects of known causes." Any result, which is the natural and regular effect of a known cause, be that cause what it may, cannot be called an accident. In this light, we are persuaded that the word is greatly misapplied to the various tragic occurrences which take place upon our railroads.
>
> Words are sometimes things; for persons, who would look upon an *accident*, as a thing to be lamented or borne with resignation, but with the production or prevention of which they had little or nothing to do, might

feel differently about it, if they considered, that "those calamities which are to individuals matter of *chance*, are to the public matter of *cause* and *effect*."[56]

Barstow's objection to the application of *accident* to railroad calamities trades on the conceptual tension at the heart of that term. Modern accidents, let us remember, are chance occurrences *and* are causally determined. They are unusual *and* natural, random on a microscale *and* regular on a macroscale. It is not clear that Barstow fully appreciates the nuances of this formulation. What is clear, though, is that he refuses to admit the accident's structuring ambivalence, its constitutive paradoxicality, into his ethico-juridical scheme, with its emphasis on the imputation of personal culpability (the laying of blame). Instead, he adamantly wants *to get rid of what is accidental*, as Hegel would have it. Whereupon this semantic sleight of hand: Barstow correlates usual or "regular" causal effects with "natural" ones and, by inverse implication, the causally *unusual* or irregular with the causally *unnatural*. Such dubious reasoning then permits him to suggest the impossibility of accidents per se in cases where causes are known or where the effects of causes are not unnatural. The governing assumption here is that the accident's basis as a "matter of chance," its troublesome fortuity, contingency, indeterminacy—its very accidentality— can be nullified or neutralized through an appeal to the natural necessity and rational intelligibility of "matter of cause and effect." Once the accident's chain of physical causation is scientifically ascertained, so goes this line of thinking, not only can blame (moral or legal responsibility) be assigned to the appropriate persons or parties, but the accident as such ceases to be—indeed, is shown to have never really *been* in the first place.

So, to restate: scientific confirmation of the naturalness and physical lawfulness of the railroad disaster's causal relations suffices to erase from that disaster any semblance of chance, to obliterate that mysterious something, that *je ne sais quoi* that makes an accident an accident. How bold and hopeful this notion is! What an ingenious conceit! But does it actually work? Will it do?

Yes and no. The cultural fantasy that underwrites accident forensics—a fantasy adumbrated in With's pioneering treatise and concentrated in Barstow's terminological objection—involves the drastic reduction of the accident's dynamic materialities, symbolic complexities, and affective intensities to the singular, purely technical question of its determinate causation, so as to ameliorate, if not the accident's initial terror, then at least its lingering horror. Put differently, the dream of accident forensics is to transmute the disruptive event of the accident (encountered as menacingly occult) into a coherent and dispassionate story of causal concatenation (greeted as reassuringly natural). To this end, devices and protocols of forensic mediation are tasked with restoring to the disaster—to its fractured temporality, to its uncertain ontology—a strict *chronology*, a precise ordering of time, along with a stable *narrativity*, a plot-like arrangement of events. Endowed ex post facto with measurable duration and intelligible succession, with an essential unity and integrity, the catastrophe thereby acquires a mundane cause and, at the same time, a meaningful reason. And, as Friedrich Nietzsche observes, "reasons bring relief."[57] In our age of forensic enlightenment, this is how the technological accident is both engaged and assuaged: how its happening is authoritatively explained, how its strangeness is made comfortably familiar, how its secret becomes officially known. This is the way that its "shocking impressions" are ideally attenuated, its nervous "apprehensions" culturally negotiated, its perpetual threat ideologically contained.

There is no denying that the forensic sciences and forensic engineering have produced, and continue to produce, remarkable results in the spheres of medicine, law enforcement, criminal justice, technology, and architecture and design, to name a conspicuous few. In these and similar arenas, forensic ideas, instruments, and procedures undoubtedly "work" (more or less, most of the time). Forensic science marshals its armamentarium to identify anonymous corpses, to establish etiologies of death and disease, to facilitate criminal arrests and convictions. Forensic engineering does the same to pinpoint causes of mechanical and structural failures, to determine responsibility, to assist in apportioning blame. Thus does that institutionally sanctioned precipitate of forensic processing—namely, forensic evidence—become socially useful as well as juridically and politically effectual. Thus do forensic ways of seeing, knowing, and believing function to mold cultural sensibilities, to make "common sense."

Yet, however much convinced by the answers furnished by forensic engineering, we moderns remain only partially reconciled to them—not quite appeased by what we have learned, not quite at peace with what we know, not quite adjusted to the "truth" of the accident. For after all the inspecting and analyzing and interpreting and explaining is done, a faint anxiety still hangs; a hint of disquiet persists. We cannot shake the uneasy suspicion that forensic logic, despite its scientific validity and its demonstrated utility, fails somehow to close the epistemological circuit. Its certainties do not totally satisfy. If it succeeds in providing a map of the accident's sequence of events, that map, detailed as it might be, nonetheless discloses only so much, takes us only so far in our understanding. A moment-by-moment account of "what really happened" tells us *how* the catastrophe unfolded, perhaps. But it does not tell us, and can never tell us, *why* it did so, nor why it harmed or otherwise affected the particular persons it did. Forensic narration penetrates the core of the accident's deep grammar, as it were, but not the enigma of its casual speech; the intricacies of its *langue* ("structural" and temporally "reversible," in the words of Claude Lévi-Strauss) but not the swerves and spontaneities of its *parole* ("statistical" and temporally "nonreversible").[58] The upshot is that forensic practice unwittingly dramatizes the limits of its own discourse, betrays the horizon of its own mastery: even when its work is completed, the element of chance is not wholly eradicated. Something of the accidental, that *residual* category, stubbornly abides. The clever attempt to impose a tidy narrative of physical causation on the accident's unruly contingency ends up reinscribing indeterminacy on another level. The drive to replace senselessness with syntax and story, to convert raw chance into representational code, fortuity into facticity, generates an unassimilable excess, an uncanny remainder (part remnant, part revenant) that resists symbolic incorporation, escapes every last rule—every rule of custom and decorum, of language, of law. Such a remainder approximates the Lacanian Real: "It is this surplus of the Real over every symbolization that functions as the object-cause of desire," avers Slavoj Žižek. "To come to terms with this surplus (or, more precisely, leftover) means to acknowledge a fundamental deadlock ('antagonism'), a kernel resisting symbolic integration-dissolution."[59] The enactment of forensic desire at once creates and conceals, denies and alibis, a supplementary space in which Giddens's "marginal forms of superstition" manage to survive and continue to antagonize. The termites never quite finish their meal; there are always leftovers.

Conceiving Forensic Media

In *The Emergence of Cinematic Time*, Mary Ann Doane maintains that nineteenth-century devices of recording and representation—particularly cinema but also photography and the phonograph—played a key role in the restructuring of time and contingency under conditions of modern capitalism. In an epoch marked by rapid industrialization and the increasing rationalization of everyday life, the motion-picture camera, with its "technological assurance of indexicality," promised to seize and render legible many of modernity's defining accidents and contingencies, including an array of nerve-racking velocities, dizzying instantaneities, breathtaking ephemeralities, and wounding instabilities.[60] "Time emerges as a problem intimately linked to the theorization of modernity as trauma or shock," writes Doane. Previously a "benign phenomenon most easily grasped by the notion of flow," time becomes "a troublesome and anxiety-producing entity that must be thought in relation to management, regulation, storage, and representation."[61]

Forensic media participate in the grand project of arresting, managing, regulating, and representing the troublesome accidentality of time in modernity. To be sure, they embody, in their own peculiar manner, the impulse to exhibit traumatizing chance, to archive suddenness and shocking contingency, to impart a semblance of order and meaning to the accident's irruptions and inflictions. In their analog incarnations (graphic, mechanical-acoustic, photographic, analog-electronic), forensic media claim a privileged relation to the real based on an epistemology of the indexical trace, which trace is regarded as the authenticating sign of presence, the guarantor of "the condition of the having-been-there," as Rosalind Krauss (following Roland Barthes) puts it.[62] In their newer, digital-electronic incarnations, forensic media deliver putative truths through the agency and cultural authority of the computer program, "hard data" through the lightning efficiencies of the algorithm and the microprocessor. Pertinent as they are, these technical attributes and epistemological ascriptions do not in themselves serve to sharply differentiate forensic-media forms and practices, as I conceive them in these pages, from other kinds of evidentiary- or documentary-media forms and practices. So, then, what distinguishes them?

I focus on two basic applications of forensic media vis-à-vis accidents of accelerated mobility, and these applications complement one another both institutionally and ideologically. On the one hand, forensic media are used in retrospect to scientifically discover the precise causes and circumstances of an unplanned high-speed crash (as were the black boxes of Air France Flight 447). On the other hand, they are used in a deliberately staged "accident" to scientifically dissect the complicated motions and concussions of fast-moving vehicles and forcefully thrown bodies (as were the overcranked cameras that recorded John Paul Stapp's rocket-sled experiments). Modern life is continually beset by injurious mishaps and ruinous breakdowns, its daily rhythms ineluctably subject to violent interruption from myriad technological collapses and failures, collisions and malfunctions. Whether in solemn response to the latest catastrophe or in anxious anticipation of the next one—present-day crash tests are, in Karen Beckman's felicitous phrase, "cinematically documented 'preenactments' of future technological disasters"—forensic media are routinely enlisted as superhuman detectives or, alternatively, to borrow Steven Shapin and Simon Schaffer's coinage, as "virtual witnesses."[63] Time and again, they are called upon to search for telltale clues, to trace hidden causal nexuses, to provide reliable evidence, to identify sources of misfortunes, to solve puzzles of physical destruction. At every alarmingly "unlucky" turn, they are entrusted and expected—by scientists, by governments, by the press and the public—to logically reconstruct and accurately recount the essence of "what went wrong."

Forensic media can be further distinguished, I submit, by the fact that their epistemic and discursive operations typically play out in and across a certain threefold temporality. Foucault, in *The Birth of the Clinic*, argues that the medical gaze of the late eighteenth century broke with tradition insofar as it observed disease "in terms of symptoms and signs."[64] Whereas corporal symptoms allowed the disease's "invariable" configuration to "show through," corporal signs indicated its course of evolution, its shifting presentation over time.[65] "The sign announces: the prognostic sign, what will happen; the anamnestic sign, what has happened; the diagnostic sign, what is now taking place."[66] An extension of the radically reorganized medical gaze of Foucault's theorization, the forensic gaze—that mode of perception allied with strategies, and actualized through technologies, of forensic mediation—is likewise disposed to read in(to) evidentiary things a semiotics annunciatory of past, present, and future. Accordingly, the science of accident forensics "sees" in three mediatized durations; its objects (are made to) "speak" in three mediatized tenses. There is the time of *recording*, the punctual "now" of the accident's occurrence. There is the time of *playback*, the variable "now" of the accident's mechanical or digital reproduction. Finally, there is the time of *imaginary projection*, the ideal "never again" of the accident's recurrence, the preemptive "later" of its forbidden repetition. As cultural technologies, forensic media articulate these temporalities of initial capture (recording), subsequent re-creation (playback), and indefinite deferral/ideological disavowal (imaginary projection). They conjure and conjoin these moments of automatic inscription (anamnesis), clinical description (diagnosis), and speculative prescription/proscription (prognosis).

Now, it may seem odd, at first glance, to talk about forensics in connection with prognostics, to speak in the same breath of such ostensibly disparate domains of knowledge, practice, and belief. The former domain, after all, is commonly understood to be directed at the past rather than the future, to be facing rearward and tracing backward, not pointing or projecting forward. Its business is that of reconstruction, and forensic reconstruction necessitates a ritualized program of vigilant return, motivated and mediated acts of looking back (retrospection) and, oftentimes, of going back (revisitation)—back, again and again, mentally or bodily, imaginatively and analytically, to the scene of the crime or accident. If their resemblance to the recursive procedures of clinical diagnosis is thus comparatively obvious, we are all the more justified in asking what these detectivist observances, these reconstructionist performances, have to do with speculative prognosis. How is it that acts of retrospection here entail those of prospection? What, exactly, does the empirical science of forensics owe to—or, asked more ominously, how is it haunted by—the conjectural arts of foreseeing, foreknowing, forecasting, foretelling?

The spectral debt in question has a long and suitably weird lineage. As Carlo Ginzburg notes in his essay "Clues: Roots of an Evidential Paradigm," prehistoric hunters and Mesopotamian diviners alike had to be skillful readers and interpreters of natural signs, expert decipherers of their material and physical environments. In order "to discover the traces of events that could not be directly experienced by the observer," both the Stone Age hunter and the Bronze Age soothsayer carried out "minute investigation[s] of even trifling matters."[67] Such traces comprised "excrement, tracks, hairs, feathers, in one case; animals' innards, drops of oil on the water, heavenly bodies, involuntary movements of the body, in the other."[68] Their vastly different historico-anthropological contexts notwithstanding, these two types of evidence, of visual, tactile, or olfactory clue—primitive-venatic and ancient-prophetic—demanded "formally identical" methods of sensing and decoding: "analyses, comparisons, classifications."[69]

Scenes of Divination, Including Haruspication, Pyromancy and Necromancy (16th century). Engraving by Hans Burgkmair. Courtesy of Bridgeman Art Library.

Ginzburg discerns in this shared modus operandi the antecedents of a distinct epistemological model or paradigm, one that fundamentally reoriented the human sciences in the late nineteenth century. Informed by medical semiotics and symptomatology, the new model laid the foundation for the aforementioned scientifico-legal complex, entering the field of art history via Giovanni Morelli, of psychology via Sigmund Freud, and of criminology, in part, via Sherlock Holmes (via his creator Arthur Conan Doyle—himself, like Morelli and Freud, a medical doctor by training). Curiously, the term *forensic* nowhere appears in either of the essay's English-translated versions.[70] Yet the "evidential paradigm," so masterfully expounded in "Clues," is forensic through and through. Indeed, the evidential paradigm is but another name for modernity's regime of forensic reason. But the critical point here is that, in Ginzburg's account of forensics' discursive formation, the modern detective is descended not only from the backtracking hunter but also from the entrails-inspecting diviner. The forensic practitioner bears the indelible stamp of the farsighted conjecturer. The reconstructionist is historically affiliated to the clue-reading occultist.

Is it any wonder, then, given these genealogical affinities and methodological correspondences, that forensic media should be imaginatively associated with futurity? That they should be bound up with intimations of tomorrow, with tendentious visions of things to come? Is it really so surprising that they have been tethered, rhetorically and rather fantastically, to the idea of progress?

Progress Revised, Accidents Redeemed

"The idea of progress," writes Robert Nisbet in his comprehensive study of the subject, "holds that mankind has advanced in the past—from some aboriginal condition of primitiveness, barbarism, or even nullity—is now advancing, and will continue to advance through the foreseeable future. In J. B. Bury's apt phrase, the idea of progress is a synthesis of the past and a prophecy of the future. It is inseparable from a sense of time flowing in unilinear fashion."[71]

In the West, the notion that history is a record of inexorable advancement from the inferior to the superior, that humankind has been improving and will continue to improve as present becomes past and future becomes present, extends back to ancient Greece. Not until the modern industrial era, however, did the idea become enmeshed with a devout belief in science and technology. "Eighteenth-century faith in progress . . . started from science; that of the nineteenth century, from mechanization," declares Sigfried Giedion in *Mechanization Takes Command*. "Industry, which brought about this mechanization with its unceasing flow of inventions, had something of the miracle that roused the fantasy of the masses."[72] Indeed, by the second half of the nineteenth century, with mechanization working its "miracle" in Western Europe and the United States, the notion of human progress—now firmly equated with *technological* progress—had, according to Bruce Mazlish and Leo Marx, "come to dominate the worldview of an entire culture."[73]

The concept of progress confronted novel and formidable challenges beginning in the early twentieth century. If, as Michael Adas contends, scientific and technological gauges of human worth and potential ruled European thought and (geo)political practice during the eighteenth and nineteenth centuries, the agonies and atrocities of the First World War (the torturous stalemate in the trenches, the murderous efficiency of the new industrialized weaponry) more than implied that those gauges were in need of recalibration, that the West's evaluation of its own moral and material preeminence was dangerously delusional, maybe even self-defeating.[74] For the conflict's mechanization of slaughter ultimately evinced not so much humanity's mastery of nature as its brutal subjugation to the very tools and techniques that emblematized and were supposed to enable that mastery. In the eyes of many moderns, the dream of technological progress had turned into a nightmare:

> Henry James's poignant expression of the sense of betrayal that Europeans felt in the early months of the war, when they realized that technical advance could lead to massive slaughter as readily as to social betterment, was elaborated upon in the years after the war by such thinkers as William Inge, who declared that the conflict had exposed the "law of inevitable progress" as a mere superstition. . . . Science had produced perhaps the ugliest of civilizations; technological marvels had been responsible for unimaginable destruction. Never again, he concluded, would there be "an opportunity for gloating over this kind of improvement." The belief in progress, the "working faith" of the West for 150 years, had been forever discredited.[75]

Giedion, reflecting in the aftermath of the Second World War, agreed: "It may well be that there are no people left, however remote, who have not lost their faith in progress. Men have become frightened by progress, changed from a hope to a menace. Faith in progress lies on the scrap heap, along with many other devaluated symbols."[76]

It has often been alleged that the idea of progress was fatally undermined two years before the outbreak of the First World War, with the sinking of the "unsinkable" Royal Mail Steamer *Titanic*. In truth, the tragedy that befell the *Ti-*

tanic was only an especially spectacular instantiation of something that by 1912 had already come to be seen as perhaps the most serious, because relentless, threat to progress: the technological accident. Referring to railroad calamities of the nineteenth century, Wolfgang Schivelbusch comments that "the accident was seen as a negative indicator of technological progress."[77] Doane amplifies the point in her classic essay on catastrophe: "The time of technological progress is always felt as linear and fundamentally irreversible—technological change is almost by definition an 'advance,' and it is extremely difficult to conceive of any movement backward, any regression. Hence, technological evolution is perceived as unflinching progress toward a total state of control over nature. If some notion of pure Progress is the utopian element in this theory of technological development, catastrophe is its dystopia, the always unexpected interruption of this forward movement."[78] "The wreck of the *Titanic* made such a tremendous impact," in Žižek's estimation, "not because of the immediate material dimensions of the catastrophe but because of its symbolic overdetermination."[79] The tragic submersion of the gigantic ocean liner presented an iconic tableau of technological regression. An accident too horrible to ignore, too devastating to discount, the *Titanic* seemed to offer startling proof of *progressus interruptus*, of forward movement "flinched."

Leo Marx's claim that "the belief in Progress has waned since it won all but universal credence within the culture of modernity" epitomizes the view of most contemporary historians and critics of the idea of progress.[80] Doubtless, today the idea, in its formulation as unstoppable, unidirectional improvement, as linear, continuous sociotechnological advancement, neither inspires the broad secular faith nor commands the profound intellectual respect that it did a century and a half ago. Nevertheless, the decline in credibility of this particular meliorist mythos, the diminished acceptance of this specific interpretation of the trajectory and telos of human history—an interpretation in which the accident constitutes "a negative indicator"—does not signal the final or absolute demise of the technoprogressivist mythology. On the contrary, as the case studies and analyses in this book collectively aim to suggest, the creed of technological progress is still very much part of the culture of modernity, albeit in significantly revised form.

In *Forensic Media*, I propose that the nineteenth-century rise of forensic rationality marked a historical mutation in the ideological relation between progress and catastrophe. Theretofore the technological accident had been considered anathema to advancement, the "wrench in the works" of civilization, the destructive factor to be eliminated, abolished once and for all, so that progress could proceed without deviation or delay. Forensics inaugurated a change in the basic terms of this equation, assigning to the presumedly negative accident a new and normatively positive use-value. To be exact, it transformed the accident into an object lesson in faulty engineering, a real-life illustration of unsafe design. As such, the accident was not just an occasion for scientific inquiry but an opportunity for practical technical instruction; not just an occurrence that government and industry learned *about* but one that engineers and designers learned *from*. Forensic discourse thereby recast dystopian catastrophe as utopian possibility. No longer "negative indicators," accidents became negative examples, were redeemed as "teachable moments." Once cursed as threats to progress, failures and disasters were now hailed as potential levers thereof. As Theodor Adorno, contradicting the elegists of progress (the Jameses, the Inges, the Giedions), stated in a 1962 lecture,

> Whoever smugly rubs their hands in remembrance of the sinking of the Titanic because the iceberg supposedly dealt the first blow to the thought of progress, forgets or suppresses that this unfortunate accident, otherwise in no way fateful, prompted measures which guarded against unplanned natural catastrophes in shipping during the following half-century. It is part of the dialectic of progress that the historical setbacks which are themselves instigated by the principle of progress—what would be more progressive than the race for the blue ribbon?—also provide the condition for humankind to find means to avoid them in the future.[81]

More recently, Paul Virilio, following Hannah Arendt, has called the accident "the hidden face of technical and scientific progress."[82]

While Adorno and Virilio each astutely perceive the necessary reciprocity of progress and catastrophe, neither ponders the historical conditionality of that reciprocity (or, what is here the same thing, the historical contingency of that necessity). For them, progress is what it is and always was, and the accident always was and is implicated—dialectically, as Adorno says—in progress's rudimentary "principle." Such a perspective is essentialist and ahistorical in its abstraction. For just as there is no immutable Nature of Accident, so there is no transcendent Law of Progress, dialectical or otherwise. Rather, there is always only "progress," always only "accident," and always only the variable historical conditions and various discursive constructions of their interrelation. Before the dawn of forensic enlightenment, progress was said to march linearly and catastrophe signified backward directionality; since then, however, the two have been forged into a kind of a negative-feedback loop (strictly speaking, the logic at work here is more cybernetic than dialectic) whereby present failures are recycled into future successes.

If practices of forensic reconstruction are the means through which accidents are produced and figured as servomechanisms of civilizational progress, then devices and protocols of forensic mediation are the means of producing and figuring those means. Forensic media—from Babbage's apparatus to aviation black boxes, from crash-test recordings to accident-scene renderings—are as much technologies of cultural imagination as they are instruments of scientific inscription, as imbued with ideological dreams as they are compelled by institutional rationales. Their first order of business might be to informationalize and anatomize the accident: to get to grips with it epistemically. And their secondary duty might be to narrativize the accident: to bring it to heel discursively and affectively. But forensic media's other, ulterior responsibility, their task in the last instance, as it were, is to fantasize the accident forward: to project its lesson, the moral of its story, into the future mythologically.

one

ENGINEERING DETECTIVES

Engineering is the art of directing the great sources of power in nature for the use and convenience of man. —THOMAS TREDGOLD, Royal Charter of the British Institution of Civil Engineers (1828)

[Engineering is] the art of moulding materials we do not really understand into shapes we cannot really analyze, so as to withstand forces we cannot really assess, in such a way that the public does not really suspect. —ERIC HUGH BROWN, *Structural Analysis, Volume 1* (1967)

In *Forensic Engineering: Learning from Failures*, a landmark collection of symposium papers published in 1986 by the American Society of Civil Engineers (ASCE), the contributor Joseph Ward defines a forensic engineer as "one who is concerned with the relationship and application of engineering facts to legal problems. He is an acknowledged expert who investigates construction-related failures and claims and subsequently determines causation and responsibility."[1] Writing fifteen years later, Randall Noon, author of *Forensic Engineering Investigation*, provides a more inclusive, and more figurative, definition of the profession and its ideal practitioner:

> Forensic engineering is the application of engineering principles and methodologies to answer questions of fact. These questions of fact are usually associated with accidents, crimes, catastrophic events, degradation of property, and various types of failures.

Initially, only the end result is known. This might be a burned-out house, damaged machinery, collapsed structure, or wrecked vehicle. From this starting point, the forensic engineer gathers evidence to "reverse engineer" how the failure occurred. Like a good journalist, a forensic engineer endeavors to determine who, what, where, when, why, and how. When a particular failure has been explained, it is said that the failure has been "reconstructed." Because of this, forensic engineers are also sometimes called reconstruction experts.[2]

Together, these congruent definitions, with their emphasis on matters of physical causation, evidentiary investigation, and failure analysis and reconstruction, lay out the basic objects and imperatives of modern forensic engineering.

The modifier *forensic* signals forensic engineering's connection to law and the legal system. Etymologically, the Latin word *forensis* (whence *forensic*) is derived from *forum*, denoting the ancient Roman "place of assembly for judicial and other public business," the equivalent of the ancient Greek agora.[3] "The forum is the arena of interpretation where claims and counterclaims have to be made on behalf of things," observe Thomas Keenan and Eyal Weizman. "And it is around disputed or contested things in particular that a forum of debate gathers."[4] In his definition above, Ward notes that the work of forensic engineering is bound up with the conduct of judicial business ("legal problems"), and, indeed, one of the forensic engineer's occasional duties is to "make claims on behalf of disputed or contested things" by offering expert testimony in court and in other legal proceedings.

In recent decades, however, the broadly public, or "agoric," aspects of forensic engineering (as opposed to the narrowly legal or judicial ones) have attained greater conspicuousness and cultural significance. "While forensic engineers continue to play an important role in litigation," remarks Kenneth Carper, editor of the aforementioned ASCE collection, "they are becoming increasingly involved in a wide variety of activities which may have little to do with litigation."[5] For example, they play an "active role in coordinating the dissemination of information resulting from failure investigation, so that design and construction procedures might be improved."[6] In this quote, Carper is referring to the dissemination of *technical* information among engineering professionals. But forensic engineers also participate, in various ways, in the *public* dissemination of forensic information, in its popular communication. They are routinely solicited, and afforded a platform and megaphone, for their expertise on how and why human-made things fall down, break down, or otherwise go amiss. They are interviewed by the press. They write mass-market books. They testify at government hearings. They are heard on radio and appear in television documentaries. Quite evidently, forensic engineers have become prominent figures and trusted institutional actors—oracles of a sort—in public debates and disclosures about the causes of mechanical and structural failures. Delivered in the voice of scientific authority, their specialized interpretations and explanations powerfully inform today's popular media representations and dominant cultural understandings of accidents and catastrophes. Their distinctive conceptions, along with the particular assumptions and ideologies that underlie those conceptions, are now part and parcel of technological "common sense."

The Paradox of Engineering Design

Probably no contemporary forensic engineer has been more publicly visible and influential than Henry Petroski. His first book, *To Engineer Is Human: The Role of Failure in Successful Design*, originally published in 1985 and republished in modified form in 1992, acquainted a sizeable lay readership with the forensic engineer's professional fixations and peculiar habits of mind. Illustrating his argument with numerous, often spectacular examples of error and disaster drawn from the ages—from the ancient Bent Pyramid to the Tacoma Narrows Bridge, from the medieval Beauvais Cathedral to the de Havilland Comet airliner and the *Challenger* space shuttle—Petroski seizes every opportunity to inculcate the principle that "the lessons learned" from technological failures "can do more to advance engineering knowledge than all the successful machines and structures in the world."[7]

This notion, that failures can be made to reveal what successes tend to conceal, that accidents even more than achievements are fonts of knowledge and spurs to advancement, Petroski names "the paradox of engineering design": "The paradox of engineering design is that successful structural concepts devolve into failures, while the colossal failures contribute to the evolution of innovative and inspiring structures. However, when we understand the principal objective of the design process as obviating failure, the paradox is resolved. For a failed structure provides a counterexample to a hypothesis and shows us incontrovertibly what cannot be done, while a structure that stands without incident often conceals whatever lessons or caveats it might hold for the next generation of engineers."[8] As "resolved," the paradox of engineering design encourages us to keep believing in our human-built surroundings, to keep trusting in the works of modern science and technology. For though our "technological lifeworld" (to borrow a term from Don Ihde) is rife with terrible breakdowns and collapses, we must remember the "very positive lesson" that "the failure of an engineering structure, tragic as it may be, need never be for naught."[9] Disasters, it turns out, are redeemable.

Since the publication of To Engineer Is Human, Petroski has continued to try to educate nonspecialist audiences in the nontechnical aspects of forensic-engineering history and practice, rehearsing or elaborating his ideas in books such as Design Paradigms: Case Histories of Error and Judgment in Engineering and Success through Failure: The Paradox of Design and To Forgive Design: Understanding Failure, in public lectures and other speaking engagements, in newspaper and magazine articles (both those written by him and those that quote him), and in radio interviews and television documentaries. Yet the gospel of "learning through failure" was spoken in engineering circles long before Petroski became its popular evangelist.

In 1924, when design errors were still something of a taboo (or at least a touchy) subject among civil engineers, Edward Godfrey self-published *Engineering Failures and Their Lessons*, an exasperated polemic against what he considered the discipline's frequent wrongheadedness regarding the causes of structural failures. In the book's introduction, Godfrey castigates the engineers of his day for their dogmatic adherence to received orthodoxy, or "established fact," in the face of demonstrably contradictory evidence.[10] He further accuses them of obscurantism and professional bad faith, owing to their defensive tendency to characterize as unforeseeable "accidents" (the scare quotes are Godfrey's) what were really disasters born of improper design.[11] Rather than take responsibility for their faulty constructions in the wake of calamity, the leaders of the engineering community instinctively circled the wagons, concocting "impossible theories that seek to place the blame of wrecks on some mysterious thing that no man could possibly anticipate and that no man ever will refer to again."[12] For Godfrey, such willful mystifications, including the self-exculpatory appeal to so-called acts of God, all but ensured that "the awful harvest of wrecks of the last few decades" would continue unabated.[13] Only by heeding the teachings of failure could the cycle of misfortune be broken: "To err is human," he concludes, but "to persist in known errors is devilish."[14]

One of the earliest documented expressions, in the modern era, of the belief in the educative value of catastrophes belongs to Robert Stephenson, son of the reputed "Father of Railways," George Stephenson. Commenting, in 1856, on a technical paper he had reviewed in his capacity as president of the British Institution of Civil Engineers, Stephenson

> expressed the hope, that all the casualties and accidents, which had occurred during their progress, would be noticed, in revising the Paper;

Railroad accident and bridge collapse, Batavia, Ohio, 1884. Courtesy of Norfolk and Western Historical Photograph Collection, Norfolk Southern Archives, Norfolk, Virginia.

for nothing was so instructive to the younger Members of the Profession, as records of accidents in large works, and of the means employed in repairing the damage. A faithful account of those accidents, and of the means by which the consequences were met, was really more valuable than a description of the most successful works. The older Engineers derived their most useful stores of experience, from the observations of those casualties which had occurred to their own and to other works, and it was most important, that they should be faithfully recorded in the Archives of the Institution.[15]

Accidents must be observed, inspected, documented, archived; they must be faithfully accounted for and dutifully learned from. Here, in essence, is Petroski's paradox of engineering design, more than a century and a quarter *avant la lettre*.

Homo Accidens

Whether announced by Stephenson in the mid-nineteenth century, polemicized by Godfrey in the early twentieth, or sloganized by Petroski in the late twentieth, forensic-engineering doctrine rests on the age-old, and fundamentally theological, notion of human fallibility. "Because man is fallible," Petroski declares, "so are his constructions."[16] Human beings, according to this enduring mythology, are essentially predisposed to error, constitutionally prone to mischance. Risky by their very nature, they build—cannot help but build—a very risky world for themselves: "Human nature in its collective and individual manifestations seems to work against achieving . . . a risk-free society."[17] Alone or en masse, people are flawed to the core, defective down to the roots: *radically* faulty. Hence, they are bound to make mistakes. They slip, they stumble, they screw up. Their plans, proverbially, often go astray. Engineering mishaps, on this view, their proliferation as well as their constant potentiality, serve as undeniable testaments to an inherently fallible humanity. "Accidents and near accidents remain our surest reminders that engineering is a human endeavor that takes place in the context of other human endeavors."[18] Ontologically distinct from beings that do not and, indeed, cannot blunder (an incomparably perfect God above, blamelessly amoral creatures below), man is uniquely the species that errs, that brings about accidents: *homo erratus*, one might say, or *homo accidens*.

Yet alongside man-the-mistake-maker stands man-the-maker; overshadowing the humbling myth of homo accidens, the heroic myth of homo faber. In the utopian forensic imagination, the bane of human fallibility is countered by the boon of human ingenuity. "As much as it is human to make mistakes, it is also human to want to avoid them."[19] In its avowed commitment to the obviation of failure, the discipline of forensic engineering lends both institutional credibility and ideological cover to what Petroski romanticizes as "the technological drive of *Homo faber* to build to ever greater heights and to bridge ever greater distances."[20] If man's reach occasionally exceeds his grasp, if his designs sometimes go awry, this natural fact is no cause for despair; on the contrary, thanks to the intelligence and dedication of the forensic engineer, it becomes a reason to rejoice. On this conception, at once a professional rationale and a technocratic rationalization, a credo and a theodicy, the nobly human propensity to design and build, when sufficiently chastened and enlightened, has the power to redress the all-too-human proneness to bungle. "To err is human," certainly. However, as the title of Petroski's best-known book triumphantly proclaims, "to engineer is human" as well.

Sick Machines and Structures

Given forensic engineering's embrace of humanist postulates and values, its heroic anthropology, it is perhaps not surprising that technological defects and wrecks have often been likened to human diseases and corpses and, correlatively, that mechanical and structural failure analysts have often been likened to medical pathologists and autopsists. As far back as 1888, a contributor to the trade journal *Engineering* offered the following analogy: "The subject of mechanical pathology is relatively as legitimate and important a study to the engineer as medical pathology is to the physician. While we expect the physician to be familiar with physiology, without pathology he would be of little use to his fellow-men, and it [is] as much within the province of the engineer to investigate causes, study symptoms, and find remedies for mechanical failures as it is 'to direct the sources of power in nature for the use and convenience of man.'"[21] In a letter to editor of *Engineering Record* printed in 1913, Godfrey, employing his usual sarcasm, protests the omission of engineering failures from the professional historical record: "When a patient is operated upon the operation is recorded in the medical journals as successful if he lives to regain consciousness. When a building is constructed it is recorded as an engineering success if it stands up until the contractor gets his money. Of course, the subsequent death of a patient or subsequent defects of a structure, if they occur, are merely incidents and have no place in professional histories."[22] After the Second World War, the disease-and-remedy metaphor was emphatically reasserted in the title and pages of Henry Lossier's *La Pathologie et Thérapeutique du Béton Armé* (*The Pathology and Therapeutics of Reinforced Concrete*), which, upon its translation into English in 1962, the eminent civil engineer Jacob Feld hailed as "landmark" text in the field.[23]

More recent examples abound, and it will suffice here to mention only a select few. James Edward Gordon, author of *Structures: Or Why Things Don't Fall Down*, writes, "All structures will be broken or destroyed in the end—just as people will die in the end. It is the purpose of medicine and engineering to postpone these occurrences for a decent interval."[24] The onetime ASCE president Joseph Ward compares forensic engineering to forensic medicine in explicit terms:

> In medicine, the forensic expert performs autopsies on human bodies to determine causation of death. When acting as a detective, the forensic medical expert may assist in the identification of those responsible for that death.
>
> In a similar way, a forensic civil engineer performs "autopsies" on bridges, buildings, dams and other engineered works to find out why they failed. If it is within the defined scope of services, the forensic engineer may render an opinion regarding responsibility for the failure.
>
> The field of forensic engineering involves many of the same challenges and rewards which are found in the field of forensic medicine.[25]

Similarly, Kenneth Carper observes that "a forensic medical investigation, although related to litigation or criminal prosecution, may result in recommendations that effect improvements in medical practice. In the same way, forensic engineers can have an impact on improved engineering design practices."[26] A contributor to the second edition of *The Engineering Handbook* regards the practice of "failure analysis as a 'postmortem' examination."[27] Last but not least, Petroski, in a piece for the journal *Technology and Culture*, talks of structures that are "sick enough to collapse" and of the engineering "physicians" whose duty is to diagnose what "ails" them.[28]

The alacrity with which forensic engineers have analogized their profession to forensic medicine is noteworthy. To speak of "pathological" machines and structures, of their "symptoms" and "remedies," their "deaths" and "postmortems," as though they were human bodies subject to morbidity and mortality, on the one hand, and to medical examination and treatment, on the other, is to speak neither purely figuratively nor merely fancifully. Quite the opposite: it is to mobilize a conceptually loaded, culturally revealing discourse. For here, as is frequently the case, a shared rhetoric and tropic vocabulary points to a common underlying rationality and sensibility. In fact, to identify correspondences between technological and anatomical things, to compare their respective forms and functions—or, more accurately, their *deformations* and *failures*—is to imply something crucial about the historical, epistemological, and even psychoemotional relations between forensic engineering and forensic pathology (the medicolegal practice of determining cause of death through examination of a corpse). More exactly, it is to suggest their symbolic affiliation and affective imbrication, to register the ideological commensurability of their fields of inquiry as well as their cross-fertilization in the cultural imagination. Through such relays and reciprocities, the two forensic disciplines exhibit what might be called a shared *manner of regard*. This manner specifies a certain operation of knowledge, vision, and power: simultaneously a particular way of evaluating or assessing something, of "sizing it up," and a particular way of seeing or beholding something, of fixing the eyes upon it. Though their domains of objects are distinct materially and institutionally, forensic engineering and forensic pathology—together with that paradigmatic forensic science known as "criminalistics," to be discussed presently—each partake of the same sociohistorical arrangement of knowledge and perception, the same culturally contingent scheme of thought and enouncement. Moreover, and accordingly, they each deploy the same sort of inspective gaze, a uniquely modern gaze—a properly *forensic* gaze.[29]

The Forensic Gaze

The forensic gaze should be understood, in part, as a discrete, latter-day articulation of what Michel Foucault, in *The Birth of the Clinic*, dubs "the anatomo-clinical gaze."[30] For Foucault, the advent of the anatomo-clinical gaze at the close of the eighteenth century marked a major turning point, a reorienting epistemological rupture in the theory and practice of Western medicine. Whereas previously the medical gaze "was directed upon the two-dimensional areas of tissues and symptoms," now it descended, three-dimensionally, penetratively, into the hidden recesses of the body—the specimenized *dead* body—in pursuit of disease's real nature, its elusive, eternal truth.[31] Rather suddenly the autopsy (from the Greek: "seeing with one's own eyes") became the primary focus of clinical experience and the privileged source of clinical knowledge.[32] The possibility of plumbing the mysteries of life and death now depended on the ability to open "up to the light of day the black coffer of the body," on the ability to expose, through carnal dissection and visual inspection, the dark, cadaverous interior.[33] With an eye recalibrated for a reconfigured epistemic field (and vice versa), medicine had entered the age of pathological anatomy, the practice of which Foucault calls "the technique of the corpse."[34]

Over time, the dissective mode of perception associated with the new medical gaze, itself a post-Renaissance instantiation of what Jonathan Sawday has termed "autoptic vision," was adapted and applied outside the purview of pathological anatomy.[35] Indeed, as it mutated and spread beyond the walls of the clinic, the strategy of anatomizing "with one's own eyes," of seeing autoptically, became an integral part of the forensic sciences in general and of criminalistics in particular.

The formal development of criminalistics in the late nineteenth century is largely attributable to the Austrian jurist and examining magistrate Hans Gross, author of *Handbuch für Untersuchungsrichter als System der Kriminalistik* (published in English as *Criminal Investigation: A Practical Handbook for Magistrates, Police Officers, and Lawyers*). Gross's handbook, which first appeared in 1893, represented a revolution in criminological philosophy and procedure, inasmuch as it attempted to modernize and rigorously scientize the methods and techniques of criminal investigation and crime reconstruction.[36] "You had only to open Gross's book to see the dawning of a new age," the criminology historian Jürgen Thorwald enthuses.

> Of course he strongly advocated anthropometry, forensic medicine, toxicology, ballistics, and serology. But in addition he had a number of chapters dealing with matters that had never been considered in a work on

criminology. Some of the chapter headings were: "Employment of the Microscope," "Employment of the Chemist," "Employment of the Physicist," "Employment of the Mineralogist, Zoologist, and Botanist." The subtitles indicated the kind of employment Gross had in mind: "Hair, Dust, Dirt on Shoes, Spots on Clothing." . . . Each of his chapters was an appeal to examining magistrates (his term for criminologists) to avail themselves of the potentialities of science and technology far more than they had so far done.[37]

Significantly, this new "police science," to use the phrase of the pioneering French criminalist Edmond Locard, himself a Gross disciple and a Sherlock Holmes devotee, was intimately connected and popularly identified with a new cultural and institutional figure, at once a new social actor and a new fictional protagonist, that forensic personage par excellence: the detective.[38]

So tightly entwined were the emergence of scientific crime detection (criminalistics) and the literary invention of the detective-hero, they demand to be spoken in the same breath—which is precisely what Ronald Thomas does in *Detective Fiction and the Rise of Forensic Science*. Thomas is keen to show how, during the nineteenth century, the latest "scientific theories of law enforcement" were reflected in the detective stories of Edgar Allan Poe, Charles Dickens, and Arthur Conan Doyle. He also shows, conversely, how those fictional tales "sometimes anticipated actual procedures in scientific police practice by offering fantasies of social control and knowledge before the actual technology to achieve either was available. . . . It became commonplace for early criminologists to attribute inspiration for their theories to the methods of a Sherlock Holmes or an Auguste Dupin."[39] Under the forensic gaze of the Holmesian or Dupinian investigator, both the body of the criminal suspect and that of the crime victim—typically a corpse—were converted into texts to be read and logically interpreted, while the criminal deeds themselves, painstakingly deduced from material traces discovered at the scene of the offense, were transposed into a linear sequence of events, an orderly chronology of causes and effects. "The systematic medicalization of crime in criminological discourse during this period corresponded to the literary detective's development into a kind of master diagnostician," Thomas writes, "an expert capable of reading the symptoms of criminal pathology in the individual body and the social body as well."[40] Besides being a preternaturally proficient reader of bodies and traces, the expert detective-diagnostician was a compelling raconteur and scrupulous reconstructionist, his task being "to tell the story of a past event that remains otherwise unknown and unexplained by fixing the identity of a suspect and fill-ing in the blanks of a broken story."[41] It fell to the forensic investigator to turn the enigmatic remains and residues of criminal trespass into a twin solution: a positive identification (answering the question "Who did it?") and a narrative explanation (answering the question "What really happened?").

Logico-temporal Reconstruction

In classical detective fiction, "the story arrives on the scene with the corpse," as Ernst Bloch says.[42] Abruptly and inexplicably *there*, the dead body raises disturbing questions, arouses an anxious curiosity. "Something is uncanny, that is how it begins."[43] Uncanny and as yet unnarrated: "Before the first word of the first chapter something happened, but no one knows what, apparently not even the narrator."[44] The corpse's unaccountable presence constitutes an intolerable mystery, a scandalous concealment; the occulted circumstances of its sudden arrival "must be brought to light."[45] Existing as an affront to reason, the unknown "'X' that precedes the beginning" must be systematically recovered and recounted, traced backward and told forward.[46] Through the meticulous collection and rational assimilation of minute clues, "those inconspicuous details so often overlooked by the constables," the gimlet-eyed detective simultaneously establishes a necessary causal nexus, relates a coherent narrative, and unmasks a reprehensible identity.[47] For Bloch, this "process of reconstruction from investigation and evidence" describes not only the detective story's "exclusive theme" but also the very form of its unfolding, the line and shape of its exposition.[48] To be sure, the hallmark of the "whodunit" (in its classical incarnation at least) is its thematization and formal recapitulation of the forensic-investigation process, its dramatization of what Dennis Porter calls "the step-by-step path of logico-temporal reconstruction."[49]

During the nineteenth century—and, indeed, ever since—the accomplishment of such a reconstruction required the professional criminalist to skillfully employ an array of scientific instruments and equipment. (Recall here Thorwald's point about the chapters in Gross's handbook.) The fictional detective, on the other hand, was required not so much to employ those technologies as to embody them, to personify them—to himself assume, sensorially and cognitively, their extraordinary indexical and archival capacities. In any case, the new science and the new literature of detection alike were, as Thomas notes, conditioned by the contemporaneous development of "practical forensic devices that extended the power of the human senses to render visible and measurable what had previously been undetectable."[50] Chief among those devices were the lie detector (polygraphy), the mug shot (anthropometric photography), and the fingerprint (dactyloscopy). Adjusted to the administrative needs of the modern police and judicial systems, and to those of the modern bureaucratic state more generally, each of these inscriptive technologies was exploited to compel the criminal body to speak for itself, to confess the truth of its shameful deviancy or secret culpability. "The jagged lines of the heart recorded by the lie detector, the lineaments of the face imprinted on a mug shot, and the swirling patterns of skin inscribed in the fingerprint all render the body as a kind of automatic writing machine."[51] By these and other state-of-the-art forensic means, both the anatomy of the criminal and the story of the criminal events—the crime's who, what, where, when, why, and how—were made newly "legible for a modern technological culture."[52]

The drive to read and methodically reconstruct a prior sequence of events, especially a sequence whose culmination engenders alarm or disquiet, underpins all forensic endeavors. As the prefix "re-" indicates, to reconstruct is to construct again, or to construct back, and something that needs constructing again, or constructing back, is something that has fallen (or been taken) apart, that has come (or been willed) to ruin or destruction. An anterior integrity has been breached: what was once whole is now broken, ruptured, shattered. Hence the moral and metaphysical overtones of the idea of reconstruction, its missionary ethos and its postlapsarian pathos. Something originally precious in this world has been damaged and must be repaired; something originally pristine has been corrupted and must be restored. A primordially sacred totality has been rent asunder, and its pieces demand reassembly. "The alpha," by which Bloch means the murder in a detective story, "happens outside of the story like the fall from grace or even the fall of the angels."[53] Reconstruction implies an ethical and spiritual vocation, carries an obligation to the fallen: as we are summoned to rescue those who have lapsed, so we are summoned to rebuild that which has collapsed. Thus do both the call to reconstruct the crime (within the realms of forensic science and detective fiction) and the call to reconstruct the accident (within the realm of forensic engineering) inevitably point back to the social, material, and symbolic violations—the alarming breaches, the disquieting ruptures—that elicited them.

Insofar as forensic reconstruction evokes multiple moments of repetition, it is haunted by the uncanniness intrinsic to all reiterations, all recursive operations. Four interrelated instances of spectral repetition, each with its own temporal structure, can be discerned.

> *First repetition*: the crime or accident is a troubling *reminiscence*. Every misdeed, every malfunction, serves as a potent reminder of other such

happenings. Today's bloody corpse, or wrecked machine or structure, bears an eerie resemblance to yesterday's: each one is new to our eyes, and we have seen them all before. "Something is uncanny, that is how it begins"—yes, but then again, it has always already begun.

Second repetition: the crime or accident suggests its potential *recurrence*. One bad turn might well portend another. Forensic reconstruction is concerned, expressly and anxiously, with preventing re-traumatization, re-wounding, the worrisome return of the same: another murder by the same perpetrator, say, or another disaster stemming from the same root. Culprits must be identified and causes ascertained "so that it will never happen again"—this the perennial forensic rationale, and battle cry.

Third repetition: the crime or accident becomes the subject of scientifically informed speculative *replay*. Mysteries are investigated through an almost obsessive-compulsive process of empirical analysis and imaginary reenactment.[54] New theories of the case are continually advanced and mentally auditioned in relation to the evidence at hand. New stories about "who did it" or "what really happened" (or, alternatively, "what went wrong") are generated, rehearsed, and revised until they "fit," until they "match" the known facts.

Fourth repetition: the crime or accident becomes an object of scientific *representation*. The peculiar art of forensic reconstruction involves numerous feats of instrumentally mediated detection and depiction. An assortment of graphic, photographic, electronic, and digital technologies—a battery of forensic media—is used to register, record, and reproduce various criminal or catastrophic phenomena in an effort to reveal their true essences.

These four spectral repetitions and their mutual interactions partly account for the aura of unease that hangs on the protocols of forensic reconstruction.

The corollary to forensic methodology's presumption of a normative causal determinism is forensic ideology's promotion of what Porter terms "the myth of the necessary chain."[55] This myth denies transgressions and technological failures even the slightest resemblance to spontaneous occurrences. Instead of as creations ex nihilo, it treats them as determinate events with discrete antecedent causes—causes naturally and unavoidably linked, serially enchained, to their effects. What is more, these concatenations of cause and effect are accorded a supreme epistemological significance, imbued as they are with the urgency of a pregnant secret and the solemnity of a sacred truth. Under regimes of forensic reconstruction, the crime's or accident's succession of occasioning circumstances is regarded as the code to be cracked, the puzzle to be pieced together. The event's chain of necessity—and what are chains if not the very emblem of stern necessity? of Ananke?—constitutes the enigmatic quarry, the elusive object of forensic desire. To come to know this chain fully and positively, to have correctly inferred each and all of its linkages, to have traced its unhappy terminus back, step by step, to its obscure *primum movens*, is to have located and captured the forensic holy grail.

Accidents of Steam and Rail

Learned interest in engineering failures and their causes is certainly not without historical precedent; notable examples can be found in classical antiquity and in the late Renaissance. Vitruvius contemplated design errors in his *Ten Books on Architecture* in the first century B.C.E., and Galileo's *Dialogues Concerning Two New Sciences*, published in 1638, "was motivated by some stories of structural failures that were inexplicable within the then-current state of the art of shipbuilding and construction practice," as Petroski observes.[56]

Yet the technical, methodological, and institutional foundations of *modern* forensic engineering—that professionalized scientific discipline to which Joseph Ward and Randall Noon refer in their respective definitions—were laid mainly in the nineteenth century.[57] They were laid, that is, during a period of rapid, profound, and far-reaching social and technological change, a time when urban environments were being rebuilt with iron and steel and when regional, national, and continental transportation infrastructures were being revolutionized by the steam engine. As a result, forensic engineering bears deeply the impress of industrialization and mechanization—or, more precisely, of industrialization *gone wrong*, of mechanization *run amok*.[58] No doubt, whatever its affinities or continuities with the design principles and construction techniques of the premodern era, the discipline of forensic engineering was decisively shaped by forms of failure and catastrophe specific to modernity—namely, not only the factory accident but also, and more important for our purposes, the steamboat accident and the railroad accident. These latter disasters proved so terrifying and bewildering, so brutally *shocking* in every sense of the word, they practically cried out for rational explanation and meaningful reassurance.

"Only a frontier massacre could match the sudden terror, the mass slaughter, the torture, and the lurid destruction attending many steamboat accidents," contends Louis Hunter, author of the classic study *Steamboats on the Western*

Steamship *Sultana* disaster. Source: *Harper's Weekly* (May 1865).

Rivers. "Here was not only an exceptional occupational hazard . . . but a hazard imposed equally upon many persons of both sexes and all ages without interest or stake in the business, the passengers."⁵⁹ Of the several classes of steamboat accident (snags, groundings, fires, sinkings, collisions, among others), none excited as much collective dread and dismay, or held as much melodramatic allure, as boiler explosions. "The unexpected suddenness and devastating force of steamboat explosions held a morbid fascination for the public, attracting greater attention and arousing more concern than other disasters on an equal and often larger scale."⁶⁰

What the boiler explosion was vis-à-vis the steamboat—intensely frightful, tremendously powerful, indiscriminately harmful, perversely sensational—the derailment was vis-à-vis the railroad train. According to Wolfgang Schivelbusch,

> The fear of derailment was ever present on train journeys in the early days. The greater the ease and speed with which the train "flew" (a typical nineteenth-century term for rail travel) the more acute the fear of catastrophe became: we have already quoted Thomas Creevy's statement made in 1829, that the railroad journey was "really flying, and it is impossible to divest yourself of the notion of instant death to all upon the least accident happening." A German text of 1845 speaks of "a certain constriction of the spirit that never quite leaves one no matter how com-

fortable the rail journeys have become." It was the fear of derailment, of catastrophe, of "not being able to influence the motion of the carriages in any way."[61]

"Death by railroad represented the first large-scale public experience with the dangers of new [industrial] technology," notes Mark Aldrich in *Death Rode the Rails*. "Early western steamboats were also risky, but fewer people experienced them.... Outside of wartime never before had such large groups of individuals been subject to such fearsome risks from manmade causes."[62]

Of course, one did not have to be an actual passenger to be alarmingly aware of the mortal dangers attending the new mechanized modes of travel or to feel something of the "spirit-constricting" helplessness indicated by Schivelbusch. Thanks to the exploitative proclivities of the commercial press, one could be so or feel so vicariously. Indeed, the popular newspapers and magazines of the day never missed the opportunity to spectacularize the latest transportation calamity (or, for that matter, the latest structural calamity: bridge disasters, too, were front-page favorites). As Robert Reed points out in "The Horrors of Travel," a chapter in his book on the history of train wrecks, "Throughout the nineteenth century America was horrified by a series of railroad catastrophes as boilers burst, bridges crumbled, and engines derailed. Wreck reports appeared frequently after 1853 in the national journals—*Harper's Weekly*, *Leslie's*, and *Ballou's*—publicizing the frightful cost of life and property. Every volume of these weekly magazines illustrated blood chilling artist's sketches of demolished passenger cars, twisted locomotives, and human debris. Daily newspapers also gave wide coverage by spreading gore across their front pages. It is no wonder that railroad accidents captured the imagination of the American public."[63] Lurid coverage of steamboat and railroad accidents during the early and mid-nineteenth century anticipated, and established a cultural template for, the press's sensationalization of electric-trolley and automobile accidents during the late nineteenth and early twentieth centuries. Together, these mass-media portrayals of risky life and grisly death in the age of accelerated mobility epitomized, in Ben Singer's phrase, "the idea that modernity had brought about a radical increase in nervous stimulation, stress, and bodily peril."[64]

Steamboats and railroads were complicated, cutting-edge technologies in this period, and the disasters that struck them were frequently difficult to fathom, even for authorities and technical experts. The origins and circumstances of boiler explosions, which plagued steamers and locomotives alike, were particularly hard to comprehend. "The causes of collisions, snaggings, and fires could be understood without great difficulty," writes Hunter, regarding steamboat mishaps. "Boiler explosions, on the other hand, were novel and extraordinary phenomena, the character and causes of which were for the man on the street shrouded in mystery and even by the scientist were only partly understood."[65] For much of the nineteenth century, scientists and mechanicians debated the subject of the root causes of boiler explosions without reaching a consensus. Several plausible theories were proffered and entertained, but none in the end proved wholly satisfactory. There were simply too many factors and variables—the machinery in question was too sophisticated, its operations too intricate and multifarious—to admit of a single, all-inclusive explanation. Hunter describes the situation in enlightening detail:

> Excessive pressure built up rapidly or slowly in a well-filled or nearly empty boiler evidently led to explosions in many cases. But so did a great many other things—a cast-iron boiler head weakened by blow holes or cracks, a badly made plate, a defective supply pump, a clogged connecting pipe, a corroded safety valve, an accumulation of mud in the boiler or a rag or broom left inside after cleaning, a poorly fastened rivet, or any one or more of a wide variety of defects in the boiler and its equipment. Inadequate safety valves, supply pumps, or supply pipes too small in proportion to the needs of the boilers, boilers of insufficient capacity for the normal requirements of the engines, or many another example of bad design and bad proportions—any one of these might produce conditions which by themselves or operating in combination with other defects might lead to an explosion. Finally there was the progressive wear and corrosion together with changes in the physical structure of the iron to which all boilers under the best of usage were subject.

Such wear and tear gradually and imperceptibly diminished the boiler's "factor of safety"—an engineering term of art denoting a structure's or system's capacity to withstand greater-than-expected loads—"and might eventually result in an explosion far more destructive than one growing out of a more obvious defect or weakness."[66]

Detour through Accident Theory

As a historically "novel and extraordinary phenomenon," the boiler explosion serves to illustrate certain theoretical conceptions about accidents that are germane to the present study.[67] Paul Virilio—whose work on the subject has become virtually canonical, the sine qua non of contemporary accident theory—has, over the course of numerous books, essays, and interviews, posited the notion of the accident's "archaeo-technological invention."[68] For Virilio, the accident constitutes a special order of supplemental creation, at once a marvel ("an inverted miracle, a secular miracle") and a manufactured thing.[69] That it is not wanted or intentionally fabricated—that, on the contrary, it arrives unbidden, all manner of pains and precautions having been taken during the design phase precisely to obviate it—is quite beside the point. The accident is a humanly engineered artifact, and it comes into being at the same time as the machine it disrupts or destroys. Though the machine alone is purposely conceived, accident and technology are born together, delivered into the world as conjoined twins, as it were. "The accident is invented at the moment the object is scientifically discovered or technically developed," Virilio declares.[70] "To invent the sailing vessel or the steam ship is to *invent the shipwreck*. To invent the train is to *invent the derailment*. To invent the private car is to produce *the motorway pile-up*. To make craft which are heavier than air fly—the aeroplane, but also the dirigible—is to invent the *plane crash*, the air disaster." On this view, a "surprise failure" is nothing more than a naive oxymoron: "As if the 'failure' were not programmed into the product from the moment of its production or implementation."[71]

Virilio's thesis, whereby every scientific or technical innovation harbors its own unique catastrophe, "its own negativity," deliberately plays on the ancient philosophical distinction between two modes of being: substance and accident.[72] The distinction goes back through Nicholas of Cusa, through medieval Scholasticism, to Aristotle, who counterposed a thing's essential (substantive) properties to its nonessential (accidental) ones.[73] The former properties, allied with necessity, are those that intrinsically make an entity what it is and without which it would cease to exist, what the Schoolmen called "quiddities." The latter, allied with contingency, are those that modify an entity in some extrinsic way but nonetheless do not change its basic nature. Essential qualities—substances—compose things, inhere in things, stand under things (sub-stance), whereas nonessential qualities—accidents—befall things, happen to things, collapse on things (*accidere*: Latin for "to fall down," also "to happen").

Virilio cleverly challenges the conventional dualism inscribed in this Aristotelian ontology, demanding "that we rethink the accepted philosophical wisdom according to which *accident* is relative and contingent and *substance* absolute and necessary."[74] The accident, he suggests, is no longer (or has never really been: Virilio's text is equivocal on this point) "accidental" in the classical sense.[75] Rather, it has become (or has always really been) substantive unto itself: "original," "primal," an immediate, irreducible constituent, a quiddity in its own right.[76] "Since the production of any 'substance' is simultaneously the production of *a typical accident*, breakdown or failure is less the deregulation of production than *the production of a specific failure*."⁷⁷ Pushing this provocation to its logical extreme, Virilio wonders, doubtless a little mischievously, if scientists and engineers might someday wish to "reverse things and directly invent the accident in order to determine the nature of the renowned 'substance' of the implicitly discovered product or mechanism, thereby avoiding the development of certain supposedly accidental catastrophes."⁷⁸

Along with the substance-accident binary, Virilio's conception draws on the other sense of "accident" in philosophical discourse, which sense likewise derives from Aristotle. In the *Physics* and the *Metaphysics*, Aristotle uses the same Greek word, *sumbebekos*, often translated into English as "accident," to denominate both a nonessential property and a chance occurrence.⁷⁹ An occurrence is accidental, for Aristotle, if it results from the unexpected, unintended confluence of two or more independent chains of cause and effect. Put differently, an accident is an event defined by the *casual* meeting or crossing of separate *causal* sequences, by the fortuitous union or intersection of mutually autonomous lines of determination. Aristotle explains:

> "Accident" means that which attaches to something and can be truly asserted, but neither of necessity nor usually, e.g.[,] if one in digging a hole for a plant found treasure. This—the finding of treasure—happens by accident to the man who digs the hole; for neither does the one come of necessity from the other or after the other, nor, if a man plants, does he usually find treasure. . . . Therefore there is no definite cause for an accident, but a chance cause, i.e.[,] an indefinite one. Going to Aegina was an accident, if the man went not in order to get there, but because he was carried out of his way by a storm or captured by pirates. The accident has happened or exists,—not in virtue of itself, however, but of something else; for the *storm* was the cause of his coming to a place for which he was not sailing, and this was Aegina.⁸⁰

Four points need to be made here with respect to this key passage from the *Metaphysics*. First, Aristotle's formulation very much accords with and supports the ordinary understanding of an accident as an unusual and unplanned event with an unforeseen cause or hidden nexus of causes. Second, the double logic of sumbebekos operative in Aristotle's doctrine—that is, the conceptual correspondence between "accident" as nonessential quality and "accident" as chance occurrence—is shown to hinge on the idea of what might be called *nonnecessary attachment*: in one sense, accidents are variable attributes that attach non-necessarily to ontologically prior entities (substances); in another, parallel sense, accidents are variable outcomes (finding buried treasure, landing in Aegina) that attach non-necessarily to temporally prior actions or intentions. Such outcomes are endings that lack compelling reasons, are ends without ultimate goals, without actual *tele*. They do not have either natural or human motivations; they do not have per se causes and do not express intrinsic purposes; they do not "exist in virtue of themselves." Accidents "simply result."[81] Accidents "just happen."

The next couple of points are especially important. Accidental events, because they have "chance" or "indefinite" rather than "definite" causes, because, according to Aristotle, they are "unstable" and devoid of inner purposiveness, represent a deviation from the laws of metaphysical causation.[82] And inasmuch as they are metaphysically deviant, inasmuch as they "swerve," they are philosophically vexing: a thorn in the side of reason, a threat to cosmic orderliness and intelligibility.[83] As the *Culture of Accidents* author Michael Witmore observes, "Aristotle's impulse is always to think about accidents as a problem, singling them out as an exception to the rules that govern change in the world."[84] While needed as part of a regimen of "epistemological hygiene," this strategy of isolating accidents and recasting them as exceptions does not suffice to quell their "disruptive power," at least not entirely.[85] Accidents have been duly identified and intellectually quarantined, to be sure, but the problem of the exception's ontological status and position, of the exact nature of its relationship to the rule—in other words, the problem, within a supposedly closed universal system, of exceptionality itself—persists. Aristotle realizes that something more radical is required. His solution? In a brazen (and dubious) move, he decrees accidents to be inherently insusceptible to rigorous speculative inquiry, unamenable to abstract thought on account of their extrinsicality and sheer contingency. He expels, anxiously, the whole category of the accidental from the realm of the authentically knowable and, in the same fell swoop, excludes fortuitous happenings from further consideration, consigning them, in effect, to philosophical irrelevance. His opening gambit: "We must first say regarding the *accidental*, that there can be no scientific treatment of it. This is confirmed by the fact that no science—practical, productive, or theoretical—troubles itself about it."[86] And then a few pages later in the *Metaphysics*, after having troubled himself about it theoretically: "Let us dismiss the accidental; for we have sufficiently determined its nature."[87] In the exposition between these peremptory assertions, between his initial depreciation and his terminal dismissal, Aristotle belittles the accidental as "practically a mere name" (the sin of nominality) and "obviously akin to non-being" (the sin of nothingness).[88] Construed as "that which is neither always nor for the most part," the accidental, in its stubborn resistance to generalization, is not so much beyond the purview as beneath the dignity of the right and proper philosopher, something only those perennial Platonic villains, the Sophists, would deem worthy of more than passing attention.[89] Plato's most famous pupil, by contrast, understands that accidents are too insubstantial and too occasional to be taken seriously, too particular to be investigated "scientifically." Philosophy is concerned to analyze and explain regularities, both celestial and terrestrial, but accidents have causes and consequences that are highly irregular. This renders them rationally unassimilable, unsystematizable. Knowledge, finally, has nothing to say about them and can do nothing with them.

Narrative, on the other hand, can speak them and does use them. Time and again in Aristotle's theoretical treatises—and this is the fourth point—accidental occurrences are clarified through the telling of stories. Consider, in addition to the tales in the *Metaphysics* of the accidentally fortunate planter and the accidentally unfortunate sailor, this example from the *Physics*: "A man is engaged in collecting subscriptions for a feast. He would have gone to such and such a place for the purpose of getting the money, if he had known. He actually went there for another purpose, and it was only accidentally that he got his money by going there."[90] When Aristotle wants to elucidate the reality of the accidental, to delineate that which "is inscrutable to man," he employs not, as he normally does, the rarified instruments of scientific abstraction but, instead, the more prosaic tools of narrative representation: character, setting, dramatic action, resolution.[91] He spins little parables about how something unpredicted took place. He relates the circumstances under which something utterly contingent unfolded in space and time, such as how a man came to collect money where and when he did not intend, how this happened not by design but *per accidens*, purely "as luck would have it." He resorts, in short, to fabulation as a modality of sense-making. For the philosopher trying to describe the accidental in all its indefiniteness, the task is to narrativize the factual conditions in play at that very moment, not to reveal the underlying laws at work in every moment. The imperative is to dramatize the punctual, not to deduce the universal. Witmore remarks "that the 'empty status' of [accidental] events with respect to per se causes is exactly what qualifies them for explanation through narrative exposition."[92] Storytelling becomes the privileged mode of explanation because "accidents happen as the result of circumstances, not rules, and only narrative can lay out these circumstances as a matter of fact rather than a function of overdetermining causes."[93]

What is more, in laying out these circumstances thus, narrative works to embed accidents within a matrix of received assumptions, expectations, and convictions, to assign them a position in an established scheme of social meanings and values. In fact, the process of narrativization is what turns ostensibly neutral, or "innocent," events into significance-laden "accidents" in the first place. Narrativization is the device that plucks them from the stream of undifferentiated things and organizes them for common understanding, thereby lending them symbolic coherence as well as ideological force. Witmore points out that accidents—in our era no less than in Aristotle's—"do not simply happen in a cultural void, but result when certain narrative conventions . . . come into contact with communal beliefs about what is likely, valuable or purposive."[94] Accidents have been the stuff of legend for thousands of years, and "a major function" of legends "has always been the attempt to explain unusual and supernatural happenings in the natural world," as the folklorist Jan Harold Brunvand says.[95] Whether told by a philosopher in the fourth century B.C.E. or popularly transmitted as "urban legends" in the twenty-first, stories about accidents are always and ultimately stories about the reciprocal dynamics of culture and cosmology.

The Crisis of Mechanical Causation

Having taken a detour through accident theory as a way to prepare the ground for the chapters that follow, let us now circle back to the issue of the boiler explosion's historical and phenomenal novelty. Created concurrently with the steamboat—and manifesting that vessel's dark underside—the boiler explosion was, to use Virilio's term, an "original accident" of nineteenth-century technoscience. The scale of its destructiveness was enormous and unprecedented, and so were the feelings of fear and anxiety it provoked in the general public.

For precisely these reasons, the boiler explosion can be read as a paradigmatic instance of Schivelbusch's principle of "the 'falling height' of a mechanical apparatus."[96] According to this principle, with its allusion to the law of gravitational acceleration, the effects of a given technological accident can be measured and evaluated along two axes of violence. Along one axis, a technology's degree of complexity is proportional to its capacity for accidental *material* damage: the more intricate or elaborate the machine, the more physically catastrophic its breakdown.[97] Along the other axis, a technology's degree of complexity is proportional to its capacity for accidental *mental*, or "spiritual," damage: the more intricate or elaborate the machine, the more harm it inflicts on consciousness, on the psyche, individual and collective, when it fails. This second axis functions as something like a "sense of security" index, a means of assessing risk affectivity. In modern industrialized societies, technology is normally lived and experienced as "second nature." People are accustomed to its omnipresence and, by and large, habituated to its smooth operation. Their subjectivities are molded by it and partly mediated through it. Their objective dealings with it are so routinized as to be automatic. The accident interrupts and unsettles these unthinking rehearsals of human-machine relations, confounds these patterns of technocultural acceptance and accommodation. "The web of perceptual and behavioral forms that came into being due to the technological construct is torn to the degree that the construct itself collapses," Schivelbusch argues. "The higher its technological level, the more denaturalized the consciousness that has become used to it, and the more destructive the collapse of both." The accident (etymologically, that which has *fallen down*, that which has *collapsed*) comes as a shock to the system, a rude awakening, a potent symptom that breaks both the promise of easy security and the spell, or thrall, of comfortable amnesia. "In the technological accident and the shock released by it, the fear that has been repressed by the improvement in technology reappears to take its revenge."[98]

The original accident of the boiler explosion, together with those other original accidents of nineteenth-century transport, the derailment and the railroad "smashup," marked the onset of an epistemological crisis peculiar to modernity: a *crisis of mechanical causation*. At the heart of this crisis resided perplexing questions about the nature of agency and of responsibility—or, in legal parlance, of liability—in a world of complicated machinery and perilous technology. Who should be held accountable when a steamboat engine bursts into flames? Who is in the wrong when a train jumps its track? How should blame be apportioned when there occurs an accident involving multiple operators and conveyances? "Just as no one theory and no one mechanical cause could be held responsible, no one human agent could be identified," writes Nan Goodman, author of *Shifting the Blame*, in reference to boiler explosions. "The candidates—the engineers, pilots, captains, managers, and owners—were numerous, and the extent of their responsibilities overlapping and diffuse."[99]

Throughout the nineteenth century, the epistemological uncertainties and ethical ambiguities surrounding mechanical causation were repeatedly examined and disputed in the law courts. Over time, a "new doctrine of accidents" emerged and gained currency.[100] Predicated on a radically transformed conception of negligence, this new doctrine replaced the old standard of "strict liability" in cases of accidental injury, loss, or damage. Under strict liability, which had prevailed in Anglo-American jurisprudence for centuries, the perpetrator of an accident was considered legally responsible regardless of whether he or she had behaved in a blameworthy manner. The question of "fault" was deemed immaterial and was not subject to adjudication. All that the plaintiff needed to demonstrate was that the mishap, or "misadventure," resulted from the defendant's actions. Such demonstrations, however, became increasingly hard to accomplish as technology—particularly transportation technology—became ever more advanced.

A spike in the number of collision-related lawsuits around the turn of the century inaugurated a shift in the basic terms and parameters of accident law in the United States. These suits, whether pertaining to crashes between horse-drawn carriages or between waterborne vessels (or, a bit later, between railroad trains), were, according to the legal historian Morton Horwitz, "the first to involve joint actors, a factor that inevitably led judges and juries beyond the simple inquiry into whether an injury had been committed in order to determine which party had 'caused' the injury."[101] And making a determination of this sort was no small feat, as Goodman suggests: "The relativity of causation and the remoteness of so many modern, mechanized injuries made it necessary to distinguish between long-standing conditions and intermittent causes, minor and substantial contributions, and initial and subsequent intervening factors in a myriad of injuries."[102] Now that cause could no longer be so readily ascertained, liability could no longer be so directly imputed. Chance impacts were precipitated by the simultaneous actions of multiple independent agents (in the mutual collision, the echo of the Aristotelian accident resounds), and suddenly it was incumbent on the courts not only to scrutinize causal sequences, which entailed the tricky business of separating "remote" from "proximate" causes, but also to decide who, in failing to exercise "ordinary care," deserved the brunt of the blame.[103] Thus was spawned "the modern concept of negligence—the concept that determines liability for accidents on the basis of an understanding of fault and carelessness."[104]

The crisis of mechanical causation that led to the formulation of a new legal doctrine of accidents was part of a broader transformation in the social experience, cultural imagination, and discursive construction of the accident during the nineteenth century. "Technology has extended and widened the notion of accident," proclaims Octavio Paz, "and what is more, it has given it an absolutely different character."[105] Indeed, with the arrival of industrial technology and mechanized travel, the accident's "character" was altered fundamentally and irreversibly. The entry for *accident* in the *Oxford English Dictionary* hints at this historical mutation: from (1) an "event," or "anything that happens" (obsolete); to (2) an "unusual event" with an "unknown cause"; to, finally, and more restrictedly, (3) an unusual event with an unknown cause that is also "unfortunate"—to wit, "a disaster." Tellingly, as an example of the latter, the *O.E.D.*

offers this usage from the year 1882: "*Daily News* 10 July 3/6 Serious railway accident: thirty persons injured."[106]

Schivelbusch notes that in Denis Diderot and Jean le Rond d'Alembert's eighteenth-century *Encyclopédie*, "accident" is only a coincidence, not a catastrophe, and certainly not a technological one. "The pre-industrial catastrophes were natural events, natural accidents. They attacked the objects they destroyed from the outside, as storms, floods, thunderbolts, and hailstones. After the Industrial Revolution, destruction by technological accident came from the inside. The technical apparatuses destroyed themselves by means of their own power. The energies tamed by the steam engine and delivered by it as regulated mechanical performance destroyed that engine itself in the case of an accident. The increasingly rapid vehicles of transportation tended to destroy themselves and each other totally, whenever they collided."[107] When the technological milieu changes, so do the cultural perceptions and cosmological implications of the accident. Beginning in the nineteenth century, catastrophes, removed from nature, remade from the inside out, are identified more and more with accidents of mechanical apparatus. The word *catastrophe* (from the Greek: an "overturning," a "sudden turn") comes to specify the devastating effects of artificial momentum ill harnessed, of technology out of control.[108] "Catastrophe does . . . always seem to have something to do with technology and its potential collapse," asserts Mary Ann Doane. "And it is also always tainted by a fascination with death—so that catastrophe might finally be defined as the conjuncture of the failure of technology and the resulting confrontation with death."[109] Often deadly in fact and always deathly in spirit (first spectral repetition: a troubling reminiscence), the accident is not an "act of God," that antiquated name for natural misfortunes ascribed to supernatural influence. Rather, it is a Frankensteinian force majeure, an abrupt and monstrous, albeit only temporary, "overturning" of the modern technological order.

From Coroners' Juries to Railway Detectives

At first, and for a long while, steamboat and railroad accident investigations, if undertaken at all, were officially the province of coroners' juries. In his study of the rhetoric of nineteenth-century accident-investigation reports, John Brockmann explains that "such juries predated the *Magna Carta* and were a 'tribunal of record' composed of twelve citizens meeting to collect evidence shortly after an accident and usually at the scene of an accident."[110] As coroners' juries were not composed of experts in the technical arts, their evidentiary findings, such as they were, did not receive the imprimatur of "science." Confronted with both the mechanical complexity of the new transportation technologies and the bureaucratic complexity of the corporate-capitalist enterprises that administered those technologies, coroners' juries were frequently overwhelmed and out of their depth. Befuddled by the diversity of forces and elements involved in a catastrophe, they often found themselves at a loss to single out causes and morally responsible actors (which, in turn, made it difficult for the courts to mete out punishment).

The critics of coroners' juries were outspoken and unsparing. In 1877, for example, the trade journal *Railroad Gazette* condemned them for their

> stupidity, and, when the subject to be investigated involves . . . the most abstruse facts . . . , or drawing correct conclusions from premises involving profound mathematical and scientific questions, it is very rare that either coroners or coroners' juries have any special training which would qualify them for investigating intelligently the causes of ordinary railroad accidents, and, when such inquiries involve some of the most profound questions of engineering science, the average coroner and his jury are as helpless and imbecilic as so many children would be in dealing with the facts or in drawing conclusions therefrom.[111]

Two years later, Charles Francis Adams Jr., chairman of the Massachusetts Railroad Commission (and brother of Henry Adams), had this to say about the work of coroners' juries in his *Notes on Railroad Accidents*: "It is absolutely sad to follow the course of these investigations, they are conducted with such an entire disregard of method and lead to such inadequate conclusions. Indeed, how could it be otherwise?—The same man never investigates two accidents, and, for the one investigation he does make, he is competent only in his own esteem."[112] The sentiment expressed by both the *Railroad Gazette* and the distinguished commissioner, and by many other steamboat and railroad authorities besides, was straightforward: the phenomena associated with mechanized-transport accidents are so alien to everyday apprehension that it is sheer folly to entrust their investigation to nonspecialists. It makes no sense to commend to members of an institution rooted in medieval custom matters so esoteric and intellectually opaque, so technically "abstruse," not to mention so grave. Only lettered gentlemen with training in engineering science could possibly be expected to penetrate the mysteries of modern catastrophes.

To America's reliance on coroners' juries, with their penchant for "ill-considered criticism" and "crude suggestions," Adams opposed "the immeasurable superiority of the system of investigation pursued in the case of railroad accidents in England."[113] In that country, he said, "a trained expert after

Railroad accident, Roanoke, Virginia, 1892. Courtesy of Norfolk and Western Historical Photograph Collection, Norfolk Southern Archives, Norfolk, Virginia.

the occurrence of each disaster visits the spot and sifts the affair to the very bottom, locating responsibility and pointing out distinctly the measures necessary to guard against its repetition" (second spectral repetition: a potential recurrence).[114] Adams here is referring to Her Majesty's Railway Inspectorate. Established in 1840 in connection with the passage that same year of the Regulation of Railways Act, the Railway Inspectorate drew its ranks from the Corps of Royal Engineers and was empowered in the name of public safety to inspect any railway in Great Britain.[115] Though the Inspectorate would have to wait until 1871 to receive the statutory authority to inquire into the causes and circumstances of accidents, it nevertheless carried out such inquiries right from the start. "The first railway accident which had the distinction of being investigated by the Railway Inspectorate occurred on 7 August 1840, three days before the Inspectorate was formed," writes Stanley Hall in *Railway Detectives*.[116]

Detectives: the title of Hall's book on the history of the Railway Inspectorate is revealing. The military engineers the Inspectorate employed as accident investigators were known as "inspecting officers," not "detectives." Those who, during the nineteenth century, actually carried the latter designation were

Engineering Detectives   57

criminal detectives—that is, "detective policemen"—not "crash detectives," as present-day transportation sleuthhounds are sometimes called.[117] Yet Hall is perfectly justified in his anachronistic usage. Why? Because the Railway Inspectorate's inspecting officers, as the first modern professional accident investigators, embodied a response to the same basic forensic impulse, as well as an adherence to the same basic forensic ideals and methods, that produced the first modern professional criminal investigators. Historically and ideologically, railway detectives (specialized reconstructionist engineers) and criminal detectives (specialized reconstructionist policemen and their private counterparts) were cut from the same cloth.

To accent this crucial point, we need to return briefly to Hans Gross and his trailblazing criminalistics handbook. Recall that Thorwald, in order to highlight the extraordinary depth and breadth of Gross's scientific commitments, cites *Criminal Investigation*'s inclusion of chapters or sections not only on anthropometry, forensic medicine, toxicology, ballistics, and serology, but also on microscopy, chemistry, physics, mineralogy, zoology, and botany. For whatever reason, Thorwald does not mention that this same handbook—a *criminalistics* handbook, let us remember—also includes a chapter, the last of twenty, on "Serious Accidents and Boiler Explosions." Of the many intriguing things about chapter 20 of *Criminal Investigation*, it will suffice here to indicate three.

First, Gross uses the boiler explosion synecdochically to stand for a whole "class of accident." He states, "The present general observations will be found equally applicable to a boiler explosion, a railway smash, a collision at sea, the collapse of a new building, or any other catastrophe of a similar nature, proceeding from preventible causes."[118] Technological explosions, smashes, collisions, collapses: this grouping is not random or arbitrary. Like the jurists, journalists, civil engineers, and, indeed, ordinary citizens of his day, Gross recognizes in these mechanical and structural failures the existence (or the invention, in the Virilioian sense) of a new and distinct—and particularly horrific—breed of disaster. "If the Investigating Officer remains rooted to the scene of the catastrophe, he is assailed on every side," Gross warns. "The destruction is generally great; dead bodies lie about; the groans of the wounded are heard everywhere; the scene of the accident presents a terrible chaos; the whole is a spectacle of desolation, disorder, and confusion."[119]

Second, Gross, apostle of positivist science that he was, insists that boiler explosions and like catastrophes must be regarded as mundane phenomena uncontaminated by any mystical element or influence. They do not attest to the interventions of fickle Fortuna; they are not attributable to the vagaries of damnable chance. "None are due to so-called bad luck or pure accident."[120]

Instead, they are rationally intelligible events with physical causes and human culprits. In a passage that conjures Edward Godfrey's critique of his fellow civil engineers, and, moreover, that resonates with the modern legal concept of negligence as elucidated by Horwitz and Goodman, Gross avers that boiler "explosions are not the effect of mysterious causes which we know not how to control; but really demonstrate the ignorance and gross carelessness of those interested." Continuing in this vein, he quotes both Adolphe Peschka, author of *Boiler Explosions and How to Prevent Them*, and a British technical report:

> "The causes," [Peschka] says, "can be discovered only when we know the antecedent circumstances, what was the condition of the boiler, the level of the water, the steam-pressure just before the accident, in fact when we know everything about the boiler and have examined its fragments." In support of this view, we may cite the report of the *Manchester Steam Boiler Association*,—"It is difficult to imagine a case that cannot be explained by natural and well-known laws, and could not have been prevented by well-known and approved methods."
>
> Hence it follows at once that in every explosion some human agency, *some man*, is at fault; any *a priori* theory as to luck or unexplainable causes must be rigorously discarded; it is also established that the cause can be determined.[121]

Here, again, we encounter a familiar line of reasoning. No matter how strange or inscrutable they initially seem, technological accidents are fully answerable to the laws of nature. Every catastrophe forms (and figures) the end of an unbroken causal chain extending backward in time. Such a chain, however complex or obscure, can be investigated empirically and reconstructed scientifically. "The cause can be determined." And once the cause is determined—through the examination of "fragments," through the disclosure of "antecedent circumstances"—the guilty party can be identified and brought to account. Once the true story of the accident is uncovered, blame can be confidently assigned. There are no good grounds to invoke "luck" or "chance" or "fortune"—or, for that matter, "accident." These notions are to be summarily rejected as so many superstitions and irrationalisms, "rigorously discarded" as so many modes of magical thinking. For this is indeed where forensic logic and procedure inexorably lead: to the refusal, or disavowal, of the accidental per se.

The third noteworthy item in the "Serious Accidents and Boiler Explosions" chapter is its mention, in passing, of accident *prevention*. "In every case what has to be done is to find the preventible cause that has not been prevented, and the person responsible for that negligence," Gross declares. "The first necessity in this connection is that the authorities should be awake to the danger and do all in their power to prevent similar accidents."[122] How, exactly, the authorities should go about achieving this precautionary "necessity" Gross does not say. But what is interesting here is the presumption of a link between accident investigation and accident prevention, the way in which success in the former endeavor is supposed to facilitate the latter. Gross takes for granted that forensic discovery—the process of finding out the natural cause ("what really happened," "what went wrong") in conjunction with the negligent person ("who did it")—can play an instrumental role in avoiding "similar accidents" in the future. He assumes that the peculiar knowledge acquired through forensic inquiry can be mobilized to limit the potential of the accident's recurrence, to forestall a certain form of spectral repetition. In short, he, like Godfrey and, before him, British Institution of Civil Engineers President Robert Stephenson, believes in the educative value of catastrophes, specifically those that have been subjected to the dissective mediations and diagnostic power of the forensic gaze.

The New Engineer-Hero

Gross's intimation that the path to a less accident-ridden future runs through the forensically inspected wreckage of the present brings us back to the gospel of learning through failure and the paradox of engineering design. From the railroad train to the Corliss engine, the skyscraper to the Brooklyn Bridge: these and other spectacular, large-scale products of mechanical and structural engineering were nothing if not wonders to behold—technologically sublime objects—during the nineteenth century.[123] But when they failed catastrophically, as they occasionally did, and when those failures were sensationalized in the media, as they routinely were, impressions of sublime wonder turned into convulsions of sublime terror. Major accidents tended to undermine faith in modern technology and to breach trust in modern engineering. That faith, that trust—in industry and applied science, in what Lewis Mumford calls "technics and civilization"—needed to be restored.[124] Explanatory appeals to "some mysterious thing that no man could possibly anticipate and that no man ever will refer to again," as Godfrey mockingly put it, were at best rationally unconvincing, at worst downright ridiculous. Engineers, not demiurges, fashioned these machines and structures, and engineers had the duty to explain their "mysterious" breakdowns.

Hence the heightened importance that came to be ascribed to accident investigation, to its clinical and methodical execution, first in Britain (as institutionalized in Her Majesty's Railway Inspectorate), then in America and throughout the Western industrialized world. In the antebellum era, coroners' juries were the norm in the United States. After the Civil War, however, "disinterested" scientific experts, as Brockmann terms them, "entered the investigations and their findings began to take more and more of a central role. Some of these 'disinterested' judges were pioneer scientists and civil engineers, while later government commissions such as those of the Massachusetts Railroad Commission under Charles Francis Adams, Jr. took center stage in the 1870s, leading eventually to the creation of the Interstate Commerce Commission in the mid-1880s."[125] During his tenure at the Massachusetts Railroad Commission, Adams championed the "disinterested" scientific—or what we would nowadays call "forensic"—approach to accident investigation (à la the Railway Inspectorate and also the Franklin Institute's Committee on Science and the Arts, located in Philadelphia).[126] In the introduction to *Notes on Railroad Accidents*, he wrote that victims of railroad disasters

> do not lose their lives without great and immediate compensating benefits to mankind. After each new "horror," as it is called, the whole world travels with an appreciable increase of safety. Both by public opinion and the courts of law the companies are held to a most rigid responsibility. The causes which led to the disaster are anxiously investigated by ingenious men, new appliances are invented, new precautions are imposed, and a greater and more watchful care is inculcated. And hence it has resulted that each year, and in obvious consequence of each fresh catastrophe, travel by rail has become safer and safer, until it has been said, and with no inconsiderable degree of truth too, that the very safest place into which a man can put himself is in the inside of a first-class railroad carriage on a train in full motion.[127]

The blatant hyperbole and dubious sincerity of the last sentence notwithstanding—after all, the bulk of Adams's book amounts to prima facie evidence *against* the proposition that motional railroad carriages, even first-class ones, are "very safe places"—this quotation distills the techno-utopian essence of the new accident calculus at the dawn of the forensic age: every apparently senseless fatality is to be reckoned as a sacrifice for the greater good; every appalling catastrophe finds a silver lining in progressive improvements in safety.

In his essay "The Anxiety of the Engineer," Ernst Bloch, whose philosophical reflections on detective fiction were earlier discussed, evokes the perpetually precarious conditions of technological modernity through the image of the metropolis suspended in midair:

The city of ever-increasing artificiality, in its detachment and distance from the natural landscape, is simultaneously so complex and so vulnerable that it is increasingly threatened by accidents to the same extent that it has rooted itself in midair—that is, the city is built upon roots that have grown more and more synthetic. This grandly suspended, inorganic metropolis must defend itself daily, hourly, against the elements as though against an enemy invasion. But most important, these elements are not of the old kind, made up of conventional modes of chance and accident. Instead, they dwell amid the complexities of mechanized existence itself; with respect to "nature," they inhabit nothingness: a nature consisting of nothing but calculations, a nature that arrived with the machine and that increasingly has taken up residence under ever less perceptible conditions, in ever more "mathematized" dimensions.[128]

The engineer, it seems, is racked with anxiety because he knows that the slightest miscalculation on his part could, and probably would, bring about a massive collapse. One tiny arithmetic error and the entire apparatus crashes and burns, the whole "synthetic" thing comes tumbling down. Yet even if every one of his calculations proves correct for a period of time, even if homo faber succeeds "for a decent interval" (to use J. E. Gordon's richly suggestive term) in keeping hidden his shameful doppelgänger homo accidens, the threat of chance destruction persists all the same. Risks do not vanish, and the engineer's dread is not attenuated, much less extinguished. For he remains acutely aware that his intricate fabrications are still vulnerable to attack—to attack from within, from the "enemy" internal to the machine itself. "In venturing forth, [the engineer] practices a good deal of cunning and draws upon tried-and-true information, equipped with safety fuses—while fearing that some fuse might prove too weak and burn out when the current is applied. The chance of an accident should rightly and properly be reduced to a minimum; yet this chance increases with every advance into the unknown."[129] It is no longer the old elemental accidents that most concern the engineer, the so-called natural disasters, those primordial and implacable forces of air, fire, earth, water. It is, rather, the newer, denaturalized accidents, those contingencies and uncertainties that "dwell amid the complexities of mechanized existence," that now weigh most heavily on his mind.

Bloch's engineer is plainly a *mechanical* engineer, a designer and builder of modern technological objects. But are these complicated artifacts of engineering really "objects" in the traditional sense? Bloch has his doubts: "Machines have been built according to such an alienated form of understanding, and pushed so far into the state of artificiality—and even partly beyond the category of objects—that they have begun to populate a new realm of the spirits."[130] *A new realm of the spirits*: the anxiety of the engineer, it turns out, is not equivalent or reducible to a simple, sublunary fear of mathematical error, to a mundane worry about the possibility of calculative mistake. No, something deeper, something weirder, more fundamentally *uncanny*, is involved. "The more advanced and unfrivolous technology is, the more mysteriously it mingles with the realm of taboo, with mists and vapors, unearthly velocity, golem-robots, and bolts of lightning. And so it comes into contact with things that were formerly conceived as belonging to the *magical sphere*."[131] Despite his professed allegiance to science, the engineer worries, perhaps secretly, perhaps only subconsciously, that he may at bottom be no more than a sorcerer. He suspects, against all good sense, that his "advances into the unknown" are underwritten by demons, or at least attended by ghosts. "*I did not call you forth*—this single phrase captures the feeling of anxiety in its highest, oldest, and strongest articulation. Congruent with this feeling of anxiety is the idea that impure spirits have somehow taken possession of a smoothly operating mechanism in order to goad it toward some hapless end."[132] Jean Baudrillard makes a similar point when he says that every accident is today interpreted as "a piece of *sabotage*. An evil demon is there to make this beautiful machine always break down. Hence [our] rationalist culture suffers, like no other, from a collective paranoia."[133]

Bloch's essay well suited the times. Written at the end of the 1920s, it used the conceit of the engineer's anxiety as a point of departure for a series of critical speculations on the ever-intensifying dangers, alienations, and phantasmic atmospherics of the technological lifeworld. More than a century's worth of mechanical and structural calamities—including the infamous sinking in 1912 of the RMS *Titanic*, to say nothing of the industrialized, chemicalized battlefield deathscapes of the First World War—had set the stage for a less-than-romantic depiction of the modern engineer (and of his modernist exploits). Long hailed as both an icon of technocratic civilization and an agent of technological progress—"a messianic figure, at least a priestly one," in Cecelia Tichi's words—the engineer is transformed, in Bloch's account, into a neurotic and a borderline paranoid, a onetime visionary now daunted by the prospect of something as ostensibly minor as a blown electrical fuse, now unnerved by something as patently irrational as the machinations of "impure spirits."[134] Previously considered a glorious incarnation of the doctrine of progress, the engineer becomes a conspicuous casualty—at once a symbolic accident and an accidental symbol—of that doctrine's early twentieth-century decline.

Still, all hope is not lost. For among the technological ruins (sign of the old engineer's shame) there emerges a new type of engineer-hero, searching for clues, gathering traces and vestiges, "sifting" the wreckage "to the very bottom," as Adams would have it: a failure analyst, a logico-temporal reconstructionist, a *forensicist*. Bloch does not appear to notice or acknowledge him, but the forensic engineer has come to the "interpretive arena" of catastrophe—to the accident forum, as it were—for the purpose of "making claims on behalf of things," *fallen* things.[135] He has come, like a news reporter, "to determine who, what, where, when, why, and how."[136] He has come, like an examining physician, to "investigate causes, study symptoms, and find remedies," to diagnose and relieve a "mechanical pathology."[137] He has come, like a backward narrator, "to tell the story of a past event that remains otherwise unknown and unexplained," to "fill in the blanks of a broken story."[138] And he has come, like a secular exorcist, to dispel "mists and vapors" and other maleficent specters. In so doing, the forensic engineer—the engineering *detective*—promises to redeem the technological disaster and, at the same time, to rehabilitate the dream of technological progress.

two

TRACINGS

It is not doubtful that the graphic form of expression will not soon substitute itself for all others, each time it acts to define a movement or a change of state, in a word a phenomenon of any sort. —ÉTIENNE-JULES MAREY, *La Méthode Graphique dans les Sciences Expérimentales*

Inscription is a form of intervention, into which new machinery continues to interpose.
—LISA GITELMAN, *Scripts, Grooves, and Writing Machine*

In 1839, the English mathematician, mechanical engineer, and "father of the computer" Charles Babbage designed, built, and tested what he called a "self-registering apparatus" for railroad trains.[1] Fastening to the internal framework of a railway carriage, the apparatus consisted of an independently suspended table, across which slowly scrolled a thousand-foot-long sheet of paper, upon which "several inking pens traced curves."[2] These curvilinear tracings recorded and represented a number of the carriage's performance parameters, including rate of speed, force of traction, and vertical, lateral, and terminal vibrations. The Babbage biographer Anthony Hyman explains:

> A powerful spring-driven clock . . . was adapted to raise and lower a special pen which made a dot on the paper every half second. Thus from the spacing of these dots the speed of the train could immediately be determined. The main inking pens and the ink feed gave Babbage a good deal of difficulty but ultimately the pens worked well, tracing out their curves,

followed closely by a roller faced with blotting-paper. The pens were connected mechanically to different parts of the carriage or to some special piece of apparatus. Thus was formed a multi-channel pen-recorder with mechanical linkage.³

What prompted Babbage to build such a strange contraption? What, in his view, were the stakes of his elaborate, self-initiated experiments? Why did he think the information encoded in those blotting-paper inscriptions was potentially useful or valuable?

Babbage answers these questions in "Railways," the twenty-fifth chapter of his autobiography, *Passages from the Life of a Philosopher*. "I have a very strong opinion that the adoption of such mechanical registrations would add greatly to the security of railway travelling," he writes, "because they would become the unerring record of facts, the incorruptible witnesses of the immediate antecedents of any catastrophe."⁴ Hyman puts it this way: "So impressed was Babbage with the value of the permanent records which he obtained with his apparatus, that he suggested that similar equipment should be installed on railway trains as a matter of routine, so that in the event of an accident it should be possible to determine the causes."⁵

The Epistemology of the Graphic Method

In the nineteenth century, instruments of mechanical inscription like Babbage's apparatus were many and sundry. Most of them, as Thomas Hankins and Robert Silverman observe, were integrally involved in the emergence of one of two scientific fields: acoustics or experimental physiology.⁶

In the field of acoustics, Thomas Young, an English physician and polymath, contrived a device in 1807 that used a small pencil, a vibrating rod, and a sheet of paper to render sonic vibrations visible and permanent. Working in Young's wake, several other researchers of the period, including the English physicist Charles Wheatstone and the German physicists Wilhelm Weber and Guillaume Wertheim, employed rods, tubes, and tuning forks, revolving discs and cylinders, steel styluses and inscriptive surfaces to preserve the otherwise ephemeral traces of sonorous bodies. Doubtless the most important mechanism associated with nineteenth-century acoustical science—apart from Thomas Edison's phonograph, which it prefigured—was Édouard-Léon Scott de Martinville's phonautograph, patented in 1857. Conceived as a means of instantly and automatically transcribing human speech, and modeled on the anatomical structure and functions of the human ear, Léon Scott's "phonautograph con-

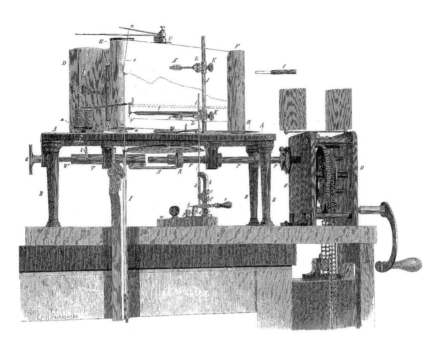

Carl Ludwig's kymograph with continuous paper. Source: Elie de Cyon, *Atlas zur Methodik der Physiologischen Experimente und Vivisectionen* (1876).

sisted of a paraboloid collecting chamber, one end of which was open, while the other was covered with a thin elastic membrane—his surrogate tympanum. The acoustic stimulation of this diaphragm activated a system of ossicle-like levers and a stylus whose motion would be traced on a steadily moving paper, wood, or glass surface coated with lampblack."[7]

In the field of experimental physiology, the German physician Carl Ludwig invented the kymograph (1847), an instrument that, when connected by catheter to the blood vessel of a dog, measured and graphically registered arterial pulse and pressure. In 1855, another German physician, Karl Vierordt, built a machine, the sphygmograph, that did the same, only non-invasively. Inspired by Ludwig's kymograph, both Carlo Matteucci, an Italian physicist, and Hermann von Helmholtz, a German physician and physicist, made use of mechanical tracers (Helmholtz dubbed his device the "myograph") in their respective experiments on muscular contractions. While the accomplishments of these researchers were considerable, no one during the second half of the nineteenth century did more to assiduously refine the procedures, ingeniously extend the applications, or tirelessly champion the diagnostic utility and scientific valid-

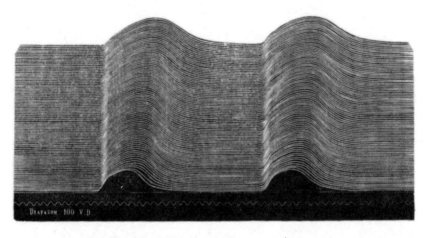

Automatic graphic inscriptions of muscular shocks. Source: Étienne-Jules Marey, *Du Mouvement dans les fonctions de la vie* (1868).

ity of automatic graphic inscription than the French physiologist Étienne-Jules Marey.

Marey was obsessed with motion in all its manifestations. An apostle of positivism, he believed that the attainment of true knowledge—by definition, *scientific* knowledge—necessitated precise observation and painstaking experiment. Accordingly, Marey made motion submit to exacting empirical investigation. His groundbreaking experiments, conducted over the course of several decades, focused on two basic kinds of movement: those internal to the animate body (heartbeat, blood flow, respiration, muscle activity) and those outwardly expressive of the animate body (walking, running, jumping, flying). The only way to scientifically know the nature of motion, according to Marey, was to measure and analyze the shifting relations of time and space as a body (or bodies) changed position(s). Before turning to photography in the early 1880s, Marey measured and analyzed the movements, internal and external, of a wide variety of animate bodies by rigorously applying the techniques and technologies of automatic inscription, an approach he famously termed "the graphic method."

Marey was the graphic method's most innovative practitioner as well as its most ardent proponent. In the 1860s alone, he invented or reinvented an astounding array of sophisticated medical-notation machines, including myographs, sphygmographs, cardiographs (heartbeat), pneumographs (respiration), thermographs (heat), and polygraphs (pulse, heartbeat, respiration,

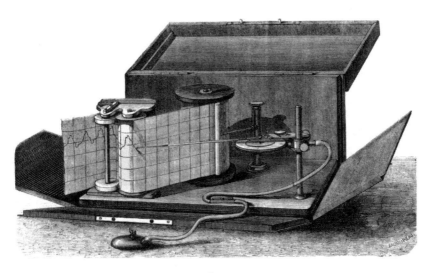

Polygraph with continuous paper. Source: Étienne-Jules Marey, *Du Mouvement dans les fonctions de la vie* (1868).

and muscular contraction). For Marey, these instruments were "indispensable intermediaries between mind and matter" because they enabled researchers to overcome "the defective capacity of our senses for discovering truths, and the insufficiency of language for expressing and transmitting those we have acquired."[8]

Combining the functions of detection, inspection, representation, and preservation in a single ostensibly objective image, mechanical-inscription devices could "see" more clearly and rapidly, "feel" more acutely and discriminatively, and "remember" more accurately and enduringly than their corresponding human modalities or faculties. What is more, they could obtain data from otherwise inaccessible areas (such as the bodily interior), translate those data into an immediately legible "language"—the "natural language of the phenomena themselves," in Marey's phrase—and store them for subsequent scientific examination and review.[9] "There is nothing that can escape the methods of analysis at our disposal," Marey declared. "Not only are these instruments sometimes destined to replace the observer, and in such circumstances to carry out their role with incontestable superiority, but they also have their own domain where nothing can replace them. When the eye ceases to see, the ear to hear, touch to feel, or indeed when our senses give deceptive appearances, these instruments are like new senses of astonishing precision."[10] Untainted by the subjectivity of the human observer, the graphic method made possible the instantaneous capture of what Marey called "the most fleeting, most delicate, and most complex movements that no language could ever express," while rendering the duration of those movements infinitely divisible (in theory, at least).[11] Robert Brain notes that, for Marey and the likeminded engineer Ernest Cheysson, "graphic recording devices appeared not merely as an effective laboratory method, they became the primary technique of universal communication, 'perfectly suited [in Cheysson's words] to the age of steam and electricity.'"[12]

These new recording technologies differed significantly from earlier incarnations such as Christopher Wren's weather clock (1663), which charted air temperature and wind direction, and James Watt's indicator diagram (1796), which plotted steam-engine cylinder pressure in relation to piston position. Pre-nineteenth-century mechanisms of this kind reduced or eliminated human labor and tedium by automating the process of accumulating data. They were appreciated because they eased research routines and made experimental procedures more efficient. Nineteenth-century graphic recording devices, by contrast, yielded the kind of data that no amount of direct observation or manual notation ever could. They were appreciated not because they reduced or eliminated something deemed undesirable (the exertions and exhaustions of the record-keeping researcher), but rather because they positively produced something otherwise unobtainable—namely, detailed knowledge of physical forces and physiological processes. They were considered revolutionary at the time because, to advisedly invoke the masculinist and ocularist metaphorics of heroic science, they penetrated, peered into, and pictured nature's secret spaces in a way theretofore impossible.

Babbage's self-registering apparatus was allied with these nature-penetrating, nature-picturing machines both technically and conceptually. All of them were intended and implemented to measure and make permanent the fugitive motions of phenomenal reality, be those motions vibratory, undulatory, oscillatory, circulatory, or locomotory. All embodied and facilitated nineteenth-century experimental science's fascination with duration and dynamic processes, with transferences of energy and their temporalities, as manifested in waves and flows, pulsations and contractions, "rhythms" (Marey's term) and "shakes" (Babbage's term). This fascination spurred the drive to discover the putative laws governing the relations among time, space, force, and movement. "All movement is the product of two factors: time and space," Marey proclaimed in 1878. "To know the movement of a body is to know the series of positions which it occupies in space during a series of successive instants."[13] Designed to register and record the various forces and vibrations of a train in motion, and to do so as a function of time and space, Babbage's apparatus, no less than Marey's battery of mechanical tracers, instantiated the cultural desire to make "the movement of a body" an object of scientific knowledge during the nineteenth century.

Besides its technical and conceptual affinities to them, Babbage's apparatus bore an *epistemological* "family resemblance" to the phonautograph, the kymograph, the sphygmograph, the myograph, and the like.[14] That is, it resembled them insofar as it evinced a foundational belief in both the scientific legitimacy and the superior reliability (as compared to unaided human senses) of the graphic method. These twin claims of legitimacy and reliability, in turn, were predicated on the epistemological promise of the inscriptive apparatus: the promise of *indexicality*. In Peircean semiotics, "index" denotes a non-arbitrary relationship between the signifier and the signified. Immediately inferable or observable, this relationship is not primarily based on either convention (as with the symbolic sign) or resemblance (as with the iconic sign). Instead, the indexical sign is causally or physically connected to its object—an object that necessarily exists "objectively," in the material world, as opposed to one that only exists "subjectively," in the mental world. For Babbage, the promise of mechanical indexicality was present in the promise of "the unerring record" and "the incorruptible witness." It was present in the promise of the technology's intrinsic impartiality and evidentiary trustworthiness: "Even the best and most unbiassed judgement ought not to be trusted when mechanical evidence can be produced."[15] This assertion, made by Babbage in reference to his railway experiments of 1839, comports exactly with Marey's claims about the graphic method's "astonishing precision" and its "incontestable superiority" vis-à-vis the sensorially defective, insufficiently expressive human observer. Philosophically and procedurally, then, Babbage was a bona fide graphic methodologist.

Whether implicit in the design of their experiments (as in Babbage's case) or explicit in the content of their writings (or both, as in Marey's case), the graphic methodologists of the nineteenth century thought it imperative that the phenomenon under investigation author its own inscriptions. "In experiments . . . which deal with time measurements," Marey wrote, "it is of immense importance that the graphic record should be *automatically* registered, in fact, that the phenomenon should give on paper its own record of duration, and of the moment of production."[16] The requirement that the motional body originate its own tracings—a requirement that represents the cornerstone of the graphic method—assumes and implies both the *automaticity* of the recording process and the *authenticity* of its products. Discursively, such assumptions and implications serve to naturalize the apparatus's operations and inscriptions, to make the machine an ahistorical, noncultural thing. Automaticity becomes a form of bodily reflex, immediate and instinctual, while authenticity becomes a function of what Léon Scott termed "natural stenography."[17] "Automatic," in other words, means that the technology works in the same manner as does an involuntary nerve impulse, while "authentic" means that its resultant texts are written in what Marey regarded as "the true universal language," the mother of all mother tongues, the lingua franca of nature.[18]

The conceit that a machine, when purposefully affixed to an organic body, is capable of imitating living tissue, that it can perform feats of what John Durham Peters calls "neurophysiological mimicry," constitutes a sort of protocyborgian mirror image of the idea, espoused by Marey and countless other scientists of the period, that a vital organism is essentially an animate machine.[19] In accordance with this perverse yet perfectly logical extension of mechanist doctrine, automaticity comes to designate the convergence of the machinic body and the organic body, while authenticity comes to designate the equivalence of mechanical stenography and "natural stenography." Nineteenth-century graphic recording devices, as Brain says, "effectively turned the body inside out, producing a freestanding, functional double of human functions."[20] Just as the graphic method made the technology responsive to stimuli through bodily incorporation (the body assimilates the machine), so it made the body communicative through technological excorporation (the machine translates the body).

Accident Anxieties

New technologies neither arrive ex nihilo nor exist in a sociohistorical vacuum, and Babbage's apparatus was, of course, no exception. Indeed, this curious recorder of railway-carriage jolts and shakes was similar technically, conceptually, and epistemologically to an array of inscriptive instruments employed in the fields of acoustics and experimental physiology during the nineteenth century.

Nevertheless, such similarities do not describe or encompass all of Babbage's apparatus's conditions of possibility. For that apparatus also reflected and expressed a quite different configuration of cultural concerns and institutional contexts—concerns and contexts that sharply distinguished it from contemporaneous applications of the graphic method. In contrast to graphic methodologists working in the acoustical and physiological sciences, Babbage devised a mechanism to detect and document, not the reverberations of nature or the rhythms of life, but the violent vibrations of technological accident, the shudders of sudden breakdown, the tremorous forces of catastrophe. The motivating question was not "How does nature work?" but rather "Why did tech-

Railroad accident, Kentish Town, England, 1861.

nology fail?" Babbage's apparatus thus crystallized the peculiar cultural anxieties and psychosocial insecurities that accompanied the arrival of the passenger railroad, haunted as it was by the specter of high-speed disaster, with all the dangers to life and limb entailed therein. It also embodied the twin impulse or tendency, incipient in Babbage's day, to rationally master and to ideologically and affectively manage the accident through forensic techniques (the analysis and interpretation of the apparatus's inscriptions) and forensic technologies (the promise and productivity of the apparatus itself).

Let us return momentarily to two passages cited in the previous chapter, the first from Robert Reed, the second from Wolfgang Schivelbusch. "Throughout the nineteenth century America was horrified by a series of railroad catastrophes as boilers burst, bridges crumbled, and engines derailed. Wreck reports appeared frequently after 1853 in the national journals—*Harper's Weekly*, *Leslie's*, and *Ballou's*—publicizing the frightful cost of life and property. Every volume of these weekly magazines illustrated blood chilling artist's sketches of demolished passenger cars, twisted locomotives, and human debris. Daily newspapers also gave wide coverage by spreading gore across their front pages. It is no wonder that railroad accidents captured the imagination of the American public."[21] Whereas Reed is interested in the sensationalization of

Tracings

train wrecks in the American popular press, Schivelbusch is concerned with the suspicions and trepidations of railway travelers in Western Europe during the same period:

> The fear of derailment was ever present on train journeys in the early days. The greater the ease and speed with which the train "flew" (a typical nineteenth-century term for rail travel) the more acute the fear of catastrophe became: we have already quoted Thomas Creevy's statement made in 1829, that the railroad journey was "really flying, and it is impossible to divest yourself of the notion of instant death to all upon the least accident happening." A German text of 1845 speaks of "a certain constriction of the spirit that never quite leaves one no matter how comfortable the rail journeys have become." It was the fear of derailment, of catastrophe, of "not being able to influence the motion of the carriages in any way."[22]

Drawing attention to the train wreck's radical egalitarianism, as it were, Roger Cooter observes that in England, Babbage's home country, railroad "accidents began deeply to concern the public from before the middle of the nineteenth century. It mattered little that there were then (in England) twice as many persons killed annually by horses and horse conveyances as by railways (1000 to 500), and six times as many drowned. The awesome 'railway dragon' had psycho-social and economic dimensions—indeed, theological and intellectual implications—which were lacking among most other potentially large-scale and routine sources of accidents, for railway accidents involved (and continually threaten to involve) persons of status and wealth."[23]

Babbage gives varied and repeated expression to these popular fears and anxieties in the "Railways" chapter of his autobiography. He communicates his concern for "the safety of all [railway] travellers."[24] He recounts his participation in numerous conversations with merchants, politicians, bureaucrats, industrialists, and other prominent personages, in which the railroad's "various difficulties and dangers were suggested and discussed."[25] He writes of "having a great wish to diminish [such] dangers" and of "studying the evidence given upon the enquiries into the various lamentable accidents which have occurred upon railways."[26] He explains the designs, functions, and purposes of railway-safety devices of his invention besides the self-registering apparatus, including one for dispatching obstructions on the tracks (which came to be called, rather grimly, a "cowcatcher") and another for preventing passenger-coach derailment in the event of engine derailment. Finally, as one who publicly participated in Britain's legendary "Battle of the Gauges"—surely one of the greatest

Railroad accident, Philadelphia, Pennsylvania, 1856.

technology-standards debates of the industrial era—he defends his early endorsement of, and continuing preference for, broad-gauge rather than narrow-gauge track, in the name of safety, economy, and efficiency.[27]

The most intriguing expression of accident anxieties, however, comes at the beginning of the chapter. There, Babbage delivers, not an exultant account of the grand and glorious railroad, as might be expected from an accomplished inventor and lifelong champion of science and industry, but instead a harrowing anecdote about an ill-omened, phantasm-inducing train ride.

Recalling his experience at the ceremonious opening of the Manchester and Liverpool Railway on 15 September 1830, Babbage writes, "We had not proceeded a mile before the whole of our trains came to a standstill without any ostensible cause." After several motionless minutes, "a certain amount of alarm now began to pervade the trains, and various conjectures were afloat of some serious accident."[28] The conjectures were soon confirmed: Member of Parliament William Huskisson, whom Babbage "had seen but a few minutes before standing at the door of the carriage conversing with the Duke of Wellington," had been accidentally run over and killed.[29] Hyman remarks that, as the first-recorded casualty in the history of railway travel, Huskisson's "death cast a shadow over the railways which was not easily lifted. . . . [It] was a reminder of the awesome power of the machinery and was to be followed by many railway deaths in fiction."[30] Babbage

Tracings 75

remembers being so shaken by the tragedy that, in its immediate aftermath, trackside scenes of salutation took on a surreal and sinister air: "For several miles before we reached our destination the sides of the railroad were crowded by a highly excited populace shouting and yelling. I feared each moment that some still greater sacrifice of life might occur from the people madly attempting to stop by their feeble arms the momentum of our enormous trains."[31] The accident: it turns a mass of high-spirited spectators into a mob of wild-eyed suicides, an extraordinary machine into an unstoppable monster, a historic maiden voyage into a veritable Grand Guignol, a communal celebration of social and technological progress into a bloody sacrifice of person and proxy of state.

Juxtaposing vivid images of excitement and fear, death and irrationality, sheer power and absolute powerlessness, Babbage's phantasm provides a glimpse into the cultural imagination of technology—and of the technological accident—in the early years of the railroad. It bespeaks a profound ambivalence toward the new mode of transportation (and, by extension, toward the new machinery of industrialization). On the one hand, there is the marvel of the machine: its breathtaking spectacle, its inexorable momentum, its astonishing enormity; on the other hand, the madness of the machine: the bedlam it engenders, the futile resistance it elicits, the brutal sacrifice it demands. Particulars of content aside, this hallucinatory expression of ambivalence toward modern transport technology is not reducible to the realm of individual psychology. It is not simply the product of one man's mind. Rather, Babbage's phantasm, considered more properly as a cultural text, as symbolic and symptomatic of the sociotechnological order, articulates key aspects of what Leo Marx terms "the technological sublime."[32]

The Technological Sublime

The technological sublime was a potent rhetoric and aesthetic during the nineteenth century, as Marx, John Kasson, and David Nye have all shown.[33] Adapting the eighteenth-century theories of Edmund Burke and Immanuel Kant to the nineteenth-century circumstances of science and industry, these authors define the technological sublime as at once a subjective experience of and a cultural discourse about modern machinery and technology, combining elements of wonder, fascination, reverence, incredulity, and dread.

Burke contrasted the essence of the beautiful to that of the sublime. Whereas the beautiful inspires "sentiments of tenderness and affection," the sublime excites "an idea of pain and danger, without being in such circumstances."[34] As "the strongest emotion which the mind is capable of feeling," the sublime specifies the complex of sensuous energies and affective intensities provoked by encounters with objects exhibiting one or more of the following characteristics: power, obscurity, vastness, magnitude, suddenness, vacuity, silence, magnificence, darkness, infinity.[35] "Astonishment," for Burke, names the emotional agitations and mental dislocations engendered by such encounters: "The passion caused by the great and sublime in nature, when those causes operate most powerfully, is Astonishment; and astonishment is that state of the soul, in which all its motions are suspended, with some degree of horror. In this case the mind is so entirely filled with its object, that it cannot entertain any other, nor by consequence reason on that object which employs it."[36] On this conception, the state of astonishment suspends the motions of reason.

Kant revised Burke's theory of the sublime in at least three fundamental respects. First, he challenged the mutual exclusivity of the beautiful and the sublime, arguing that because pleasure is produced by sublime objects no less than by beautiful ones, the two experiences are not wholly antithetical. Yet neither are they wholly identical. Instead, beautiful pleasure and sublime pleasure are inversely charged: the former is positive; the latter, tinged as it is with terror, is negative. Because "the mind is alternately attracted and repelled by the object," says Kant, "the satisfaction in the sublime implies not so much positive pleasure as wonder or reverential awe, and may be called a negative pleasure."[37]

In his second revision to Burkean theory, Kant divided the experience of the sublime into two discrete categories: the mathematical sublime and the dynamic sublime. The mathematical sublime denotes the encounter with prodigious size or scale, overwhelming enormity or immensity, incomparable vastness or greatness. The dynamic sublime, by contrast, designates the subject's confrontation with nature's menacing features, implacable forces, or devastating furies. Kant offers a few examples:

> Bold, overhanging and as it were threatening cliffs, masses of cloud piled up in the heavens and alive with lightning and peals of thunder, volcanoes in all their destructive force, hurricanes bearing destruction in their path, the boundless ocean in the fury of a tempest, the lofty waterfall of a mighty river; these by their tremendous force dwarf our power of resistance into insignificance. But we are all the more attracted by their aspect the more fearful they are, when we are in a state of security; and we at once pronounce them sublime, because they call out unwonted strength of soul and reveal in us a power of resistance of an entirely different kind, which gives us courage to measure ourselves against the apparent omnipotence of nature.[38]

Like Burke, Kant claimed that the sublime experience is characterized by a peculiar kind of fear: a double-coded fear, a feeling of fear coupled with an awareness of the absence of danger, a fear that in Sartrean terms is simultaneously pre-reflective and reflective, a real fear that nonetheless knows not to be really afraid. Unlike Burke, however—and here is his third revision—Kant maintained that this fear ultimately mutates, by means of an operation interior to the sovereign subject, into its opposite: courage. Naked before nature, whether instantiated as infinite expanse (mathematics) or as infinite energy (dynamics), the subject initially and instinctively cowers, flooded as he is with feelings of impotence and insignificance. But then, according to Kant, the subject experiences an awakening from within, a reinforcement from reason, which, by reaching out to grasp the totality and potency of things—of sublime things—restores his sense of moral strength and superiority. Reason, infinitely more powerful than the powerful infinity it apprehends, comes to the rescue.

For Burke and Kant alike, the sublime object was necessarily a natural one (or a supernatural one: God). By the turn of the nineteenth century, however, the sublime's (super)naturalness had become secularized and assimilated to the sphere of everyday life, owing to the increasing omnipresence and, in some cases, seeming omnipotence of technology. Indeed, as Kasson notes, the forces of industrialization caused the sublime to become enmeshed with the machine in cultural imagery and imagination: "In the second half of the eighteenth century in England, . . . artists soon broadened the field of the sublime beyond the natural landscape and discovered new sources of sublime emotion in industrial processes. By the mid-nineteenth century, both in England and America, this aesthetic of the technological sublime had achieved a broad following, as the reactions to the Corliss engine dramatically illustrate. The desire to 'see sublimity . . . in the magnificent totality of the great Corliss engine,' to shiver with awe and dread before it, had become a popular passion."[39]

Those who wished to "shiver with awe and dread" before George Corliss's colossal dynamo had to visit Machinery Hall at the Philadelphia Centennial Exposition of 1876. For most citizens of industrialized nations, however, a much easier way to experience the frisson of the technological sublime was to stand alongside the railroad tracks and take in the train as it rushed by—just like the "highly excited populace" in Babbage's phantasm. "Perhaps the most common vehicle of the technological sublime in the nineteenth century was the railroad," Kasson observes. "Viewers not only thrilled to the elaborate ornamentation of locomotives such as the 'America,' which reached its height around midcentury; they were still more fascinated by the sight of the railroad in motion. Universally accessible, a rushing train possessed almost all the Burkean attributes and symbolized the beneficent new technological order. Observed at close range, particularly at night, it possessed an irresistible appeal."[40]

Kasson's point about the train symbolizing "the beneficent new technological order" is significant because it highlights the link between the rhetoric/aesthetic of technological sublime and the ideology of progress. Leo Marx contends that "the idea that history is a record of more or less continuous progress had become popular during the eighteenth century, but chiefly among the educated. Associated with achievements of Newtonian mechanics, the idea remained abstract and relatively inaccessible. But with rapid industrialization, the notion of progress became palpable; 'improvements' were visible to everyone. During the nineteenth century, accordingly, the awe and reverence once reserved for the Deity and later bestowed upon the visible landscape is directed toward technology or, rather, the technological conquest of matter."[41] Marx here indicates three historically successive articulations of sublimity: from fear and trembling before God, to fear and trembling before Nature, to fear and trembling before the Machine, conqueror and re-creator of the material world. In the last articulation, the sublime is subsumed into a modern progressivist mythos, in which feelings of "awe and reverence" for technology affirm the reality of human achievement and advancement. Hence, the technological sublime, unlike either the theological or the natural sublime, is actually more than a rhetoric and aesthetic. It is also a mode of historical consciousness: both a way of seeing the past as revealed by and redeemed in the technological present, and a way of imagining the present as prelude to and prophetic of a still more sublime technological future.

As a popular nineteenth-century "structure of feeling," the technological sublime informed not only the discursive and affective *content* of Babbage's phantasm (its mood, its imagery, its symbolism), but also the discursive and affective *context* of his apparatus's invention.[42] To be sure, the sublimity of the railroad train, its status as an object of public fascination and fear during the nineteenth century, figured Babbage's hallucinatory story as a symptomatic text, even as it formed the background against which his strange contraption emerged as a symptomatic technology. A material reminder of the awful possibility of the accident, of the deadly perils of riding the rails—and, even more ominously, of the utter *unpredictability* of those perils—Babbage's graphic recording device could not but conjure the technological sublime's aspect of dread and terror, of "pain and danger."

The Triumph of Reason and the Ideology of Progress

Babbage's apparatus conjured two other aspects of the technological sublime as well: the triumph of reason and the ideology of progress.

The Kantian sublime consists of two qualitatively distinct moments. At first, the subject, seized by something bigger, loses his bearings. Stunned by the sublime object's seemingly unfathomable infinity, ungraspable alterity, and unassimilable exteriority, his rational faculties fail him; he is perfectly mortified and practically immobilized. But then, at the moment of his maximal vulnerability, reason returns with a vengeance. The subject suddenly comes to his senses, comes to grips with the situation, for he now realizes that nature's sublime forces are no match for his powers of intellect. Secure in his ability to outthink that which initially terrorized him, he reclaims his dignity and his sovereignty—and, with them, his sense of certainty, the certainty of truth.

Beneath the design of Babbage's apparatus resides a version of this story in which reason triumphs over mortal terror. Impelling both the parable of the Kantian sublime and the intended purpose of Babbage's apparatus is essentially the same cultural mythology. For Kant, the subject is shaken to the core and shocked into submission by something that is initially beyond the bounds of his comprehension. By the end of the parable, however, the incapacitating effects of astonishing nature are neutralized by a rationality that transcends them: reason is reinthroned. The same basic storyline is built into Babbage's apparatus. The specifics of the scenario are different, of course, and the setting has shifted from the plane of the psychological to that of the technocultural. But the structure of the plot, the essence of the conflict, the identity of the hero, and the moral of the story all conform to the Kantian parable.

The drama unfolds in three acts. *Act One*: The locomotive steams down the line, its iron grip on the rails providing stability and security. It moves with great momentum in a forward direction. It is a marvel of modern science and engineering, a testament to the power of reason and a token of moral and material progress. *Act Two*: Disaster strikes! The train derails, collides, explodes, or breaks down. Its forward motion is abruptly, and violently, interrupted. The accident unleashes the forces of chaos. All around is destruction, disorder, death; all of a sudden, a scene of unreason. *Act Three*: Railway experts search the wreckage for evidence as to the cause of the accident. Babbage's apparatus is retrieved and its scrolls are scrutinized. Across their paper surface, mechanical tracings encode the tale—the truth—of the train wreck. The cause of the accident is thereby ascertained, and this new knowledge is instrumentalized to make targeted technical improvements. With order now restored and the public reassured, the marvelous locomotive once again steams down the line, even more stably and securely than before, with great momentum, in a forward direction.

This three-act drama is obviously contrived, but it serves to illustrate an important point. At the heart of every technology lies a cultural ideal and fantasy, a vision of what the technology is good at and good for, an imagining of how human objectives will be achieved and the conditions of human life improved through its agency and application. Babbage's apparatus points to the ever-present possibilities of human error and technological failure. While its manner of operation is tied to the accident's unpredictability (disaster strikes unexpectedly, so the mechanical pens must inscribe continuously), its social and economic usefulness is bound up with the accident's inevitability (disaster will strike eventually, so practical measures must be taken immediately). In fact, the accident *has to* happen, *has to* take place, in order for the apparatus to fulfill a valuable function; otherwise, or until such time, it is virtually useless. Its efficacies and efficiencies are inversely related to those of the conveyance that carries it, because it gains recognizable worth only, and exactly, when that conveyance's recognized worth is destroyed. Its potential for machinic production depends on the greater machine's sudden ruination. As an instrument of human intention, it "makes sense" only in connection with the coming to pass of a catastrophe of which it is presumed capable of "making sense." This twofold sense-making—its comprehensibility as a purposeful artifact (its intentionality) and its purported ability to comprehend the accident (its instrumentality)—implies what lies at the romantic heart of Babbage's apparatus: the dream of retroactively rationalizing the "accidental" irruption of the irrational.

Act Three portrays this process of retroactive rationalization, which is driven by a dual imperative. First, the accident must be *written faithfully* (indexically, automatically, authentically); second, it must be *read forensically*. To read the accident forensically is to perform a certain kind of scientific exegesis. It is to painstakingly analyze and rigorously interpret what is simultaneously a book of revelation (about "what really happened") and a book of judgment (about "what went wrong"). Such a book is at once open and esoteric: open because, per the graphic methodologists, it is inscribed in the universal language of nature; esoteric because those "natural" inscriptions can only be revealed—indeed, can only be registered in the first place—by means of specialized techniques and technologies. To read the book of accident is to retrace a prior sequence of events, to recover an anterior course of action, and, through this retracing and recovering, to render the awful culmination of that sequence both retrospectively inevitable and prospectively innocuous. It is to make the accident less threatening in actuality through forensic decipherment and less threatening in the abstract through forensic decipherability. To read the book of accident is to undertake to exorcise the demons of chance and contingency from the official account of "what really happened," of "what went wrong." In a broader sense, however, to read the book of accident is to undertake to exorcise those demons from *all* official accounts of technological failures and disasters, past, present, and future. For if accidents can be rationally explained, so goes the logic, they are not "accidents" at all; rather, they are events with determinable and determinate causes. To read the accident forensically, then, is to write not only the authorized history of a particular accident but also the prolegomena to any future history of accidents—a history neatly expunged of messy indeterminacy, a history of accidents immaculately evacuated of true accidentality.

Both the Kantian sublime and Babbage's apparatus tell the tale of a sophisticated and settled rationality in Act One (the sovereign subject, the railway system), the terrifying unsettling of that rationality in Act Two (the sublime encounter, the train wreck), and the reprisal and triumphant resettling of that rationality in Act Three (the subject's transcendence, the accident's legibility and intelligibility). In each case, a dread-inducing disruption to a normative rationality is suffered and finally surmounted—surmounted by the very rationality whose integrity and authority had been assailed and imperiled. The third act manifests the crucial utopian moment, executes Babbage's apparatus's *transfiguring vision*. Though, from one perspective, a stark testament to fallibility and failure, the self-registering apparatus nevertheless promises the future transcendence, if not of human frailty, then at least of technological deficiency. It promises not just to recuperate the accident in the name of reason but, further, to *redeem* it in the name of progress. "The moment of redemption, however secularized, cannot be erased from the concept [of progress]," declares Theodor Adorno. "For the enlightened moment in it, which terminates in reconciliation with nature by calming nature's terror, is sibling to the enlightened moment of nature domination."[43] At bottom, Babbage's apparatus is about the disciplined application of the instruments of rational enlightenment to overcome the disaster of rational enlightenment's own device. It is about, to borrow Ernst Jünger's trope, the "drying up" and de-realization of danger, about the perceptual reduction of apparent senselessness to understandable erroneousness:

> The supreme power through which the bourgeois sees security guaranteed is reason. The closer he finds himself to the center of reason, the more the dark shadows in which danger conceals itself disperse, and

the ideal condition which it is the task of progress to achieve consists of the world domination of reason through which the wellsprings of the dangerous are not merely to be minimized but ultimately to be dried up altogether. The dangerous reveals itself in the light of reason to be senseless and relinquishes its claim on reality. In this world all depends on the perception of the dangerous as the senseless; then in the same moment it is overcome, it appears in the mirror of reason as an error.[44]

Above, I insisted that to read the accident forensically is to read it as a book of revelation and, at the same time, as a book of judgment. Here let us acknowledge a third book of forensic imagination, after the epistemological and the moral: that of *prophecy*, the ideological book, the properly *prescriptive* book. Why this book, too? Because, "in the mirror of reason," whose reflections perforce distort even as they illuminate, discovering "what really happened" and deciding "what went wrong" are not so much ends in themselves as means *to* an end—an end that finds its prophetic fulfillment in the future. Reading forensically facilitates the construction of a "better mousetrap" (better because safer) and, by extension, of a "brighter tomorrow" (brighter because more secure, less slippery). This technoprogressivist fantasy penetrates to the core of Babbage's apparatus, as it does to those of all forensic media, intended as they are to help modern inventors, engineers, designers, and other technical experts "learn from failures."

Recall Babbage's summary statement regarding the purpose of his apparatus: "I have a very strong opinion that the adoption of such mechanical registrations would add greatly to the security of railway travelling, because they would become the unerring record of facts, the incorruptible witnesses of the immediate antecedents of any catastrophe." The device that produced those registrations, that inscribed that allegedly factual record, was fundamentally different from other railway-safety devices that Babbage invented or espoused. Whereas the cowcatcher, the anti-derailment mechanism, and broad-gauge track were all designed to protect railroad passengers in and of the present (actual passengers, riding right now), Babbage's apparatus was designed to protect railroad passengers in and of the future (abstract passengers, riding later on). Each of the first three devices, once installed, promised to provide safety immediately and more or less independently. The latter device, by contrast, held no such promise. In fact, without the substantial expenditure of industrial and institutional resources to serve and support it—including the time, labor, capital, and expertise needed to retrieve its mechanical tracings, to analyze and interpret them, and to implement the systemic changes they prescribed—Babbage's apparatus was not a railway-safety device in any meaningful sense of the term *safety*. But *with* the expenditure of those resources, it became a tool for projecting "dried-up dangerousness" into the future, an instrument for passing security on to posterity—in the end, a machine for the sublime production of social and technological progress.

Fixing the Accident through Forensic Media

In his essay "Crash (Speed as Engine of Individuation)," Jeffrey Schnapp shows that, contrary to conventional wisdom, popular anxieties concerning the speed and safety of new modes of transportation antedate the rise of the railroad. By the mid-eighteenth century, for example, the horse-drawn coach, whether a two-wheel cabriolet or a four-wheel phaeton, though a source of exhilaration for the privileged driver, had become one of exasperation for the general public. Schnapp writes, "The spread of this 'epidemic' of mobile self-display, thanks to which individuality became identified with administration of one's own speed, terrified rural populations and transformed cities like Paris and London into nightmarish killing fields in the eyes of many observers."⁴⁵

Single Harness Phaeton Horse "Columbine" (circa 1880). Lithograph by Vincent Brooks, Day and Son.

The Comforts of a Cabriolet! or, the Advantages of Driving Hoodwink'd!! (1821). Etching by George Cruikshank. Courtesy of Jeffrey T. Schnapp.

Schnapp's essay throws brilliant light on an often overlooked chapter in the historical "anthropology of speed and thrill," demonstrating that the modern cult of accelerated mobility—and trepidations about its mortal dangers—began to crystallize a century or so earlier than is usually supposed.[46] Missing from mention, however, are some historically and anthropologically noteworthy differences between eighteenth-century coach accidents and nineteenth-century train accidents. As the phrase "nightmarish killing fields" suggests, the private coach represented a hostile incursion into public space because it threatened the health and welfare (not to mention the rights and rights-of-way) of pedestrians, bystanders, and others deprived of the new technology of accelerated mobility. Of course, fears of being run over and killed by a high-speed conveyance did not fail to greet the arrival of the railroad, as the horrified public reaction (Babbage's phantasm included) to William Huskisson's untimely demise attests. On balance, however, the rails posed less of a bodily threat to those who lacked accelerated mobility than to those who enjoyed its prerogatives. If the coach primarily endangered innocents in the streets (persons and publics *outside* the vehicle), the train mainly imperiled its own occupants (persons and publics *inside* the vehicle).

Moreover, railroad accidents differed from coach accidents in their destructive intensity and magnitude—and in their psychological toll. For the traveler of the nineteenth century, the horror of the train wreck was the horror of riding (or of imagining oneself riding) the train as it wrecked. Whereas the coachman "administered his own speed" and so experienced an intoxicating sense of confidence and control, the railway passenger could not "influence the motion of the carriages in any way" and so experienced a mortifying sense of vulnerability and incapacity. "A feeling of safety is joined to the technology upon which it is based," according to Schivelbusch.

> Technology has created an artificial environment which people become used to as second nature. If the technological base collapses, the feeling of habituation and security collapses with it. What we called the "falling height" of technological constructs (destructivity of accident proportionate to technical level of construct) can also be applied to the human consequences of the technological accident. The web of perceptual and behavioral forms that came into being due to the technological construct is torn to the degree that the construct itself collapses. The higher its technological level, the more denaturalized the consciousness that has become used to it, and the more destructive the collapse of both.[47]

The railroad train, perched at the apex of technological complexity, towering above all previous modes of transportation, including the horse-drawn coach, was disposed to "fall" or "collapse" with unprecedented force and fury. So, too, was "the feeling of habituation and security" that attended it.

During the 1830s and 1840s, Great Britain and the United States, each in the throes of industrialization, saw the flowering of an extraordinary faith in the principles and products of science, industry, and technology. This faith raised "the creed of progress" into a "dogma," in Sigfried Giedion's words.[48] "Eighteenth-century faith in progress . . . started from science; that of the nineteenth century, from mechanization. Industry, which brought about this mechanization with its unceasing flow of inventions, had something of the miracle that roused the fantasy of the masses."[49] No nineteenth-century mechanized invention more impressively symbolized the creed of progress, or more vigorously "roused the fantasy of the masses," than did the railroad. Yet, at the same time, as Alan Trachtenberg observes, "the railroad was never free of some note of menace, some undercurrent of fear. The popular images of the 'mechanical horse' manifest fear in the very act of seeming to bury it in a domesticating metaphor: fear of displacement of familiar nature by a fire-snorting machine with its own internal source of power."[50] Despite its popular appellation, the "mechanical horse" was not a docile or domesticated animal. Despite the continuities between eighteenth-century and nineteenth-century "anthropologies of speed and thrill," the railway journey was experientially and affectively unlike a ride in a cabriolet or a phaeton. And despite the fact that both the coach accident and the train accident were transportation mishaps, the former neither enacted the trauma nor emblematized the tragedy of modern accelerated mobility as strikingly—as astonishingly—as the latter.

Devised and experimentally deployed in 1839—one short year before the establishment, by a British Act of Parliament, of Her Majesty's Railway Inspectorate—Babbage's self-registering apparatus at once reflected and attempted to reconcile the opposing sides of the train's technological sublimity: its masterful harnessing of the forces of nature versus its disastrous inability to completely contain them; its status as an icon of reason and enlightenment versus its reputation for error and erratic behavior; its ideology of progress versus its spectacle of catastrophe.[51] Babbage's invention announced the imperfection of the railroad's technical achievement, even as it promised that that achievement was perfectible, that it could and would be perfected in the future, so long as the accident was made amenable to devices and protocols of forensic mediation.

This orientation toward the (perfectible) future is significant. In contrast to the "nowness" of other nineteenth-century railway-safety devices, Babbage's apparatus mediated and articulated three separate temporalities: the accident in the present, the accident from the past, and the non-accident of the future; writing the accident as it happens, reading the accident as it happened, and erasing the accident's accidentality so that it will never happen again; a present-tense inscription (the language of nature), a past-tense description (a book of revelation, a book of judgment), and a future-tense prescription (a book of prophecy). This curious feature—its triple-tensing of the accident—is what distinguished Babbage's apparatus from the single-tense safety devices of its day (the cowcatcher, the anti-derailment mechanism, broad-gauge track). It is also what made it a prototypal forensic medium—to be precise, a precursor to "black-box" recorders.

Bringing together, for the first time, a medium of recording and representation, a means of accelerated mobility, and the methods of experimental science, Babbage's strange, state-of-the-art contraption embodied the twin desire to make the accident available to knowledge and susceptible to control. It marked the emergence of the modern impulse—an impulse still very much with us—to render perceptible, decodable, and explicable the physical dynamics of inauspicious chance, to fix the causal forces and relations of ruinous contingency, where "to fix" means to seize/stabilize *and* to repair/reconstruct. Bent on transmuting the dross of technological failure into the gold of rational explanation, it sought to assuage the uncertainties and insecurities surrounding new forms and practices of accelerated mobility by deciphering and disciplining transportation disaster.

Crucially, Babbage's apparatus belonged to neither acoustical nor physiological science, and it was not intended to transcribe "normal" manifestations of motion. Instead, it belonged to the prehistory of a field and profession that would not take definitive shape until the twentieth century—namely, forensic engineering—and it was intended to transcribe the decidedly "abnormal" motions of accident and catastrophe. In this last respect, it resonated with two other nineteenth-century instruments designed to mechanically register the spatiotemporal displacements and motional perturbations of the "errant," the "aberrant," the "abnormal": the seismograph and the lie detector.[52] Like Babbage's apparatus, each of these devices employed the graphic method in the hope of making an extreme social, material, or ideological disturbance (an earthquake, a violent criminal) speak its secret truth and, in so doing, surrender its terrifying power. For this reason, Babbage's apparatus can be understood as part of a broader, unmistakably modern constellation of forensic logics, tendencies, languages, technologies, and imaginings that came into being over the course of the nineteenth century, and that continues to inform and animate a wide range of cultural and institutional discourses and practices in the twenty-first.

So what became of Babbage's apparatus after 1839? As certain as Babbage was that his invention, if widely adopted, would make the railroad safer for the traveling public, he was equally certain that its wide adoption was unlikely "unless directors can be convinced that the knowledge derived from [it] would . . . considerably diminish the repairs and working expenses both of the engine and of the rail."[53] The railway system was a large-scale capitalist enterprise, and Babbage suspected that the economic value of his little graphic recording device would not be immediately apparent to those charged with managing and maintaining the profitability of that enterprise. Compounding the problem, according to Babbage, was his lack of professional credentials: "Since the long series of experiments I made in 1839, I have had no experience either official or professional upon the subject. My opinions, therefore, must be taken only at what they are worth, and will probably be regarded as the dreams of an amateur."[54] As it turned out, Babbage's vision of a widely adopted accident-encoding apparatus remained something of an "amateur's dream" for more than a century.[55]

three

BLACK BOXES

> Given a black thing, an obscure process, or a confused cloud of signals—what we shall soon call a problem. We intervene to illuminate it, define it, reduce it to something simple. Someone comes . . . [and] opens the black box, Pandora's box with all its gifts. Attracted by such a source, some others join the first, organize the work site, bringing light, equipment, documentation, increasing sophistication of means and the ever more complex organization of their group.
> —MICHEL SERRES, *The Parasite*

> What remains of people is what media can store and communicate.
> —FRIEDRICH KITTLER, *Gramophone, Film, Typewriter*

As a cultural trope and symbol, the black box is mysterious, an obscure object of desire. Hinting of menace, its dark geometries draw us in. We want to know what is inside, but we cannot really *see*. Our curiosity weirdly compels us. Can we open it, this austere yet alluring thing? Do we dare? We want to see what is inside, but can we ever really *know*?

In the scientific arena, the term *black box* gained currency shortly after the Second World War. Norbert Wiener, in his pioneering treatise on cybernetics, defined a black box as "a piece of apparatus . . . which performs a definite operation on the present and past of the input potential, but for which we do not necessarily have any information of the structure by which this operation is performed."[1] In this now-classic formulation, a black box is a machine whose inputs and outputs are specified and understood but whose internal workings are concealed and unknown. More recently, Michel Serres and Bruno Latour have borrowed and adapted Wiener's definition for their own respective projects.[2] For each of these theorists, a black box is necessarily a Pandora's box, one

Dislocation of Intimacy (1998). Art installation by Ken Goldberg in collaboration with Bob Farzin. Courtesy of Ken Goldberg.

that, according to Latour, bears an ominous inscription: "DANGER: DO NOT OPEN."³

Outside the realms of cybernetics and science studies, the figure of the black box has featured again and again in cinema, literature, and popular culture. In 1914, for example, the silent-film actor and director Norval MacGregor made a short comedy for Selig Polyscope Company titled *The Mysterious Black Box*. The following year, Universal Film Manufacturing Company released *The Black Box*, the concluding episode of a fifteen-part serial chronicling the adventures of the aptly surnamed criminalist Sanford Quest.⁴ Though the film is presumed lost, the E. Phillips Oppenheim book on which it was based survives.

In the book *The Black Box*, published in 1915 and illustrated with stills from the movie, the titular object presents a foreboding enigma:

> Quest had paused suddenly in front of an oak sideboard which stood against the wall. Occupying a position upon it of some prominence was a small black box, whose presence there seemed to him unfamiliar. Laura came over to his side and looked at it also in puzzled fashion. . . .

Newspaper advertisement for Universal Film Manufacturing Company's *The Black Box* (1915).

> Quest grunted.
>
> "H'm! No one else has been in the room, and it hasn't been empty for more than ten minutes," he remarked. "Well, let's see what's inside, any way."
>
> "Just be careful, Mr. Quest," Laura advised. "I don't get that box at all."
>
> Quest pushed it with his forefinger.
>
> "No bomb inside, any way," he remarked. "Here goes!"
>
> He lifted off the lid. There was nothing in the interior but a sheet of paper folded up. Quest smoothed it out with his hand. They all leaned over and read the following words, written in an obviously disguised hand.[5]

The black box functions as a receptacle for the depositing of—and a vehicle for the delivery of—cryptic messages. Contained inside it are strange inscriptions that must be deciphered if the identity of the murderer is to be discovered. "The mystery of the black box" demands of the master detective "the exercise of all his ingenuity."[6]

This eerie scene, which marks Quest's initial encounter with the black box (several more such encounters follow), appears in a chapter titled "The Pocket Wireless." In this chapter's narration, at the moment right before the criminalist chances upon the box, we are pointedly reminded that he possesses his own "private telegraph and cable."[7]

> Mr. Sanford Quest sat in his favourite easy-chair, . . . his attention riveted upon a small instrument which he was supporting upon his knee. . . . He glanced across the room to where Lenora was bending over her desk.
>
> "We've done it this time, young woman," he declared triumphantly. "It's all O.K., working like a little peach."
>
> Lenora rose and came towards him. She glanced at the instrument which Quest was fitting into a small leather case.
>
> "Is that the pocket wireless?"
>
> He nodded. . . .
>
> "We've got it tuned to a shade now," Quest declared. "Equipped with this simple little device, you can speak to me from anywhere up to ten or a dozen miles."[8]

Quest is here portrayed as an "early adopter" of the new technology of wireless, able to receive and decode the "speech" of distant persons from the comfort of his armchair. Moments later, he retrieves and forensically reads a missive "written in an obviously disguised hand." He is a sort of cryptanalyst, a solver

Film still from *The Black Box* (1915), directed by Otis Turner. Source: E. Phillips Oppenheim, *The Black Box* (1915).

QUEST AND LENORA RECEIVE THE MESSAGE FROM LAURA.

of cryptographs. And what are these? "Cryptography and cryptoanalysis," Friedrich Kittler tells us, are the names taken by "writing and reading under the conditions of high technology."[9]

By the time Oppenheim wrote *The Black Box*, the conventional association between forensics and cryptography was already well established, having been introduced in the second quarter of the previous century by no less an authority on both subjects than Edgar Allan Poe. As Shawn James Rosenheim observes, "the form of the detective story—invented single-handedly by Poe—is predicated on the application of a cryptographic technique to the opaque materiality of the world."[10] Quest's interest in and adeptness at wireless telegraphy are significant in this connection. "For millennia, cryptography existed in a kind of Masonic silence, in which knowledge of the art was confined to a tiny class of governmental practitioners or to those few who employed it for amusement," Rosenheim explains. "With the spread of the telegraph in the 1840s, however, this pattern began to change, as cryptography worked its way into not only the hardware of civilization but into our imaginations as well."[11] Forensics, cryptography, telegraphy: they are all of an imaginative piece. But what distinguishes Oppenheim's detective tale—and what makes it so intriguing for our purposes—is its added element, the fourth turn of the metonymic screw: forensic science, cryptanalytic "ratiocination" (Poe's term for detectivist reasoning), a modern communications medium, *and a message-bearing black box*.

In 1917, two years after the publication of Oppenheim's book, Eugene O'Neill produced *In the Zone*, a one-act "sea play" in which a sailor possessing a small, sinister-looking black box is suspected (wrongly, it turns out) by his fellow seamen of being a German spy. Some O'Neill scholars have suggested that the playwright might well have drawn inspiration for this work from short stories by Poe ("The Oblong Box") and Arthur Conan Doyle ("That Little Square Box")—those literary progenitors of forensic logic.[12]

Black boxes, moreover, turn up in the history of modern media and communications. A recurrent source of mystery and anxiety, of both wonder and dread, the photographic camera is perhaps the archetypal media-technological "black box." Let us consider—besides the fact that Eastman Kodak's "Brownie" models were often referred to as "little black boxes" during the early twentieth century, including in the company's own promotional literature—this description from the popular magazine the *Mentor* in 1928:[13]

> Like the magic carpet of the Arabian Nights tales, it whisks you anywhere and everywhere. This is black-box magic but worked with no abracadabra, no aid of a djinnie [sic] conjured forth from a bottle, lamp or ring. Just the modern magic of a little dark box, a roll of light-sensitive film, a disk of glass and a clicking shutter. But because of it any small schoolboy knows more today about what this earth is like than the wisest of the Greek philosophers. Little could Daguerre and Niépce, fellow inventors of photography a century ago, have foreseen the world of wonders that they were to reveal to us through the peephole of this little black box![14]

Much could be made of this passage. Suffice it here to note that its rhetoric at once denies and depends for its effect on the equivalence of "black-box magic" and black-box technology, that it invokes photography's exoticness and uncanniness in order to advance a claim about the camera's scientificity and epistemological superiority. "Abracadabra," we are assured, is (un)certainly (not) at work.

Guglielmo Marconi and his black box, circa 1897. Courtesy of Smithsonian Libraries.

Probably the single most famous historical episode involving a media-technological black box, however, concerns Italian inventor Guglielmo Marconi's eventful arrival in England near the end of the nineteenth century. Marconi designed and built his black box—a device for the transmission and reception of radio signals—when he was twenty-two years old. According to Erik Barnouw, Marconi's wireless prototype failed to attract the interest of the Italian minister of post and telegraph. In response to the disappointing news, the precocious tinkerer's "mother made a quick, remarkable decision: she and Guglielmo would take the invention to England."[15] "Early in 1896 young Marconi carefully packed and locked the black box that held the invention, and they set sail. In England the black box at once stirred the suspicion of customs officials. Two years earlier the French President had been killed by an Italian anarchist. In the black box were wires, batteries, and tubes with metal filings. Smashing it seemed wisest, and this they did; Guglielmo had to begin his English visit by reconstructing his invention."[16] The black box invites suspicion, induces unease. What is this weird contraption? A dynamo? A motor? A *bomb*? (Remember that Quest, too, worried about the possibility of an incendiary ex-

Crash-damaged flight recorder. Photograph by and courtesy of Jeffrey Milstein.

plosion.) The black box poses a riddle: the customs officials cannot make heads or tails of the bizarre apparatus, and they do not understand, or simply do not believe, the foreigner's technical explanation. The black box represents a threat to political stability and social order. It must be destroyed.

At the turn of the twentieth century, it was relatively easy for the English authorities to smash Marconi's black box. Today, it is practically impossible to destroy the twin aviation technologies known colloquially as "black boxes": the flight-data recorder and the cockpit voice recorder. The former device electronically or digitally records information about the airplane's in-flight performance and operating conditions, including such parameters as time, altitude, airspeed, heading, vertical and lateral acceleration, pitch and roll attitude, wing-flap movement and position, thrust for each engine, and flight-control surfaces. The latter device electromagnetically or digitally records the voices of the flight crew, automated warnings and weather briefings, radio communications with air traffic control, and ambient sounds captured by the cockpit-area microphone, including engine noise and landing-gear extension and retraction. Both the flight-data recorder and the cockpit voice recorder are equipped with "underwater locator beacons" and are housed in thermally insulated, corrosion-resistant titanium or stainless-steel cases, installed in the most "crash-survivable" part of the airframe, typically the tail section. According to Dennis Grossi, National Resource Specialist for Flight Data Recorders at the United States National Transportation Safety Board (NTSB), the purpose of these crash-protected black boxes—which, despite their common appellation, are painted bright orange for enhanced visibility—is "to provide accident investigators with the information needed to determine the cause of a mishap and take the proper corrective action to prevent a similar mishap from recurring."[17]

Early Flight Recorders

In 1937, the *New York Times* and *Science* magazine each reported on the development of a new aviation technology: the flight recorder, or, alternatively, "the flight analyzer."[18] Introduced by engineers of the U.S. Bureau of Air Commerce during a three-day air-safety conference summoned by the Department of Commerce, the mechanism consisted of a barometric altimeter, three automatic indelible-ink pens, and a three-by-five-inch card attached to a revolving cylinder, all encased in a three-pound, six-by-eight-inch duralumin shell. One pen traced the plane's altitude continuously from takeoff to landing; one registered the time and number of radiotelephone communications from pilot to ground stations; and one recorded information about the operation of the automatic pilot.[19] Installed in the tail section prior to takeoff and removed for inspection "immediately upon landing," the flight recorder was designed to function for eight hours at a stretch, which was "much longer than any scheduled non-stop flight" in 1937.[20] Its intended purpose was to improve the safety and efficiency of commercial aviation by making it "possible for the pilot himself, the dispatchers and the chief pilots to reconstruct completely the story of the plane's flight."[21]

Framing the issue in terms of pilot consciousness and conscientiousness, the *Times* stressed the device's utility as an instrument of discipline and deterrence:

> For a long time operators and government officials have been trying to devise ways and means to make the pilots "altitude conscious." Low flying, they believe, has been responsible for a number of serious accidents, and the accident reports of the bureau have pointed this out. While there is a government regulation requiring an altitude of at least 500 feet above all ground obstacles in the best of weather and higher flying when the weather is bad, there hitherto has been no way of checking exactly how

high the planes stay, save when a disastrous crash reveals the fact that the pilot has been too close to the ground. "This instrument should make the pilots 'altitude conscious,'" one official said today, "and should go a long way toward making flying safer."[22]

Science also commented on the recorder's surveillance capacities: "Company officials will also be able to check on whether safety regulations with regard to altitudes at which the planes fly have been carefully observed. A check is also provided on the airways' traffic control scheme enforced in major airline services." At the end of the piece, a rather more ominous potential use for the automatic-writing machine was mentioned: "The analyzer will also aid future safety work by providing a permanent record of what went on in the plane before any accident that might occur. Analysis of accident causes has frequently been hampered in the past by the fact that little was known of the plane's behavior immediately before disaster overtook it."[23] "To *reconstruct* completely the *story*" of the "disastrous crash." To "aid *future safety* work by providing a *permanent record* of what went" wrong. The "*analysis* of accident *causes*." Here, already (and again): the discourse, the dream of forensic mediation.

In 1937, *Science* reported that sixty United Air Lines planes had been equipped with the flight analyzer, and that all commercial airliners in the U.S. would be so equipped in the near future. Instead, the near future brought the Second World War, the exigencies of which demanded an emphatic reordering of military and civilian priorities, along with a strategic reallocation of social, economic, and technological resources.

During the war, flight recorders of one kind or another were implicated in the problem of national security through their role in military flight-testing. *Aviation* magazine announced in 1942 that "a corps of electronic and mechanical design engineers" at Brown Instrument Company, a subsidiary of Minneapolis-Honeywell Regulator Company, had developed "a new type of self-balancing electronic potentiometer" (an instrument for measuring electromotive forces) specifically for the Douglas B-19, then the largest airplane in operation.[24] The device, which automatically generated a graph-paper printout of bulkhead, wing-strut, and tail-surface pressures, as well as motor, carburetor, exhaust, and oil-line temperatures, promised to relieve the flight-test engineer of his manual note-taking duties. "With an instrument of this type installed in the plane," wrote *Aviation*, "readings will be made regularly and automatically and the pilot may devote his entire attention to the proper manipulation of the airplane controls. Another advantage of the flight recorder is that many more

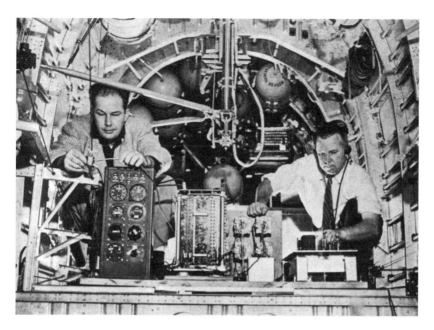

Installing the Vultee Radio-Recorder's onboard transmission unit. Source: *Radio News* (February 1945).

readings may be obtained in a given time interval so that much more complete and detailed information of the flight is available than has heretofore been the case where readings were recorded manually."[25]

During the same period, Vultee Aircraft Corporation manufactured and marketed its own flight recorder. More sophisticated than Brown Instrument's potentiometer, the Vultee Radio-Recorder, or "test-flight stenographer," electronically sensed a wide range of strains, motions, pressures, and temperatures by means of specially engineered pickups. In that same instant, it radioed (the technical term is *telemetered*) the data to a ground station, where it was recorded simultaneously on several different media: wax disks, sound film, and paper charts.[26] Like the Brown Instrument recorder, the Vultee recorder was designed to improve, through automation, the reliability and efficiency of flight-data collection. Both devices worked to remove the human (slow, distractible, limited, fallible) from the process. Only the test-flight stenographer, however, worked to reduce the inscriptive apparatus's exposure to accidental destruction.

Because its sensing mechanism was located some distance from its record-

The Vultee Radio-Recorder's ground-station receiving apparatus. Source: *Flying* (May 1943).

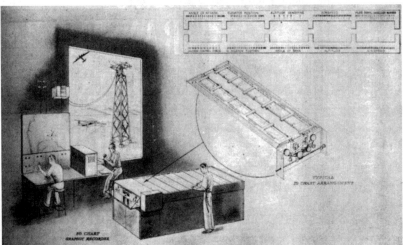

The Vultee Radio-Recorder's ground-station components. Source: *Radio News* (February 1945).

ing mechanism, the one in the air, the other on the ground, the Vultee Radio-Recorder's transcriptions would remain undamaged in the event of a flight accident (unless, of course, the plane crashed into the ground station—an improbable circumstance, to be sure). *Radio News* highlighted this aspect in 1943:

> Pilots testing warplanes are faced with the necessity of controlling their planes during severe gyrations, and cannot be expected to record faithfully the many instrumental readings. Because the many factors of atmospheric conditions, vibrations, fuel flow, etc., are often variables, frequently changing by split-seconds, readings and recordings must be taken very rapidly. One reason alone is adequate to provide radio transmission for these readings. Should the plane crash, valuable data would otherwise be lost. What caused the failure? Very likely we never would learn the answer. As it is, however, those records are safely in the keeping of engineers on the ground at the very moment those in the plane are lost.[27]

Two years later, another *Radio News* article implied the test-flight stenographer's links to both accident forensics and technological progress: "[The Vultee Radio-Recorder gives] us the composite picture of the airplane, instantaneously and accurately. Once we relied upon motion pictures and 'magic eye' cameras to record instrument indications. If an experimental airplane crashed, odds favored the records being mutilated. Now, thanks to electronic recording, we can know what took place up to the instant of the crash, and perhaps find from those records how to avoid a repetition. In any event, electronic recording marks great progress in flight-testing of aircraft."[28] The motion-picture medium was all but defenseless. Whether stored in cockpit or cabin, film lay unguarded, unprotected, its photochemical materiality vulnerable to aviation catastrophe, its corpus subject to "mutilation." Telemetry, on the other hand, "instantaneous and accurate," promised the post-crash survivability of flight data—data that could be used precisely to repel the specter of disastrous repetition.

As it happened, interest in radiotelemetric flight recorders waned in the postwar years, perhaps because of the various costs and complexities involved in building, operating, and maintaining the ground-station components. In contrast, electromechanical flight recorders, after a wartime hiatus, became objects of considerable technical innovation, commercial investment, and governmental attention and regulation.

In June 1940, the Air Safety Board, which had been created two years earlier with the passage of the Civil Aeronautics Act, was abolished at the behest of President Franklin D. Roosevelt, and its accident-investigating duties were transferred to the newly established Civil Aeronautics Board (CAB).[29] Responding to a spate of airliner crashes in the early 1940s, the CAB called for the installation of some kind of onboard automatic record-keeping mechanism that would remain intact in the event of an accident: a crash-protected flight recorder. Wartime shortages continuously delayed the development of a such a mechanism, however, and, "after extending the compliance date three times, the CAB rescinded the requirement in 1944."[30]

Three years later, in September, the CAB revisited the issue, adopting Civil Air Regulation amendments that required that flight recorders be installed, by the end of June 1948, on all cargo and passenger planes. In March 1948, the requirement was relaxed so as to make it "applicable only to planes with a certificated maximum takeoff weight of 10,000 lb. or over."[31] A few months later, it was rescinded altogether. *Aviation Week* covered the story in July 1948: "The Board's Safety Bureau now has announced that delays in producing adequate flight recorders made it impossible for the airlines to comply with the requirement by June 30. It also said that because all of the flight recorders available are of new design, some operational experience is desirable to prove the serviceability and dependability of each type before requiring that all transport aircraft be equipped with the device."[32] Vacillations in government policy throughout the 1940s nonetheless did not dissuade some manufacturers from attempting to answer the CAB's call for a state-of-the-art flight recorder.

Science Digest reported in 1947 that General Electric Company had been working on an "automatic flight recorder, designed to provide recorded data which will help determine the cause of aircraft mishaps."[33] The GE recorder had two distinctive characteristics. First, it registered the plane's altitude, airspeed, compass heading, vertical acceleration, and other parameters by means of an inkless mechanism: a stylus etched a groove, two-hundredths of an inch wide, into black paper coated with a thin layer of white lacquer, thereby obviating the difficulties that sometimes vexed inking instruments at high altitudes (blotting, clogging, freezing).[34] Second, it used a system of tiny electrical signaling devices, or "selsyns," to instantly transmit readings from the plane's instruments to the onboard flight recorder. Engineers at GE claimed that their "new instrument promotes safety . . . by furnishing recorded data which may later determine the cause of an accident to the plane."[35] Of course, the usefulness of such data depended on the robustness of the medium that stored it and—crucially, in the case of flight recorders—on the sturdiness of the container that surrounded it. Yet, apart from the fact that it was made to be placed in the airplane's tail section, General Electric's automatic flight recorder possessed little in the way of actual crash protection.

James Ryan (left) and the General Mills Ryan Flight Recorder, 1953. Courtesy of the General Mills Archives.

The General Mills Ryan Flight Recorder

Several years later, a seemingly less likely corporate "General"—General Mills, the Midwestern cereal company—debuted its own flight-data recorder.[36] The General Mills Ryan Flight Recorder, so named for the mechanical engineer James J. Ryan, the company employee and University of Minnesota professor credited with its invention, was designed to provide what *Flying* magazine, in 1954, called "an absolute and permanent record of an aircraft's air speed, altitude, vertical acceleration, time and heading."[37]

Self-contained save for its connections to the plane's gyrocompass, General Mills's flight recorder, unlike General Electric's, was entirely mechanical, requiring no internal electronics or remote pickups. Driven by a battery-powered spring motor and capable of operating for up to three hundred hours without reloading, its oscillographic apparatus fed a sheet of aluminum foil, one millimeter thick and two-and-a-quarter inches wide, through an escapement mechanism that controlled its rate of speed, between three-and-a-half and five-and-a-half inches per hour. Each of four styluses inscribed a discrete data parameter: velocity, g-force, altitude, time. (An early incarnation of the device bore the name Ryan VGA Flight Recorder, the initials "VGA" standing for the first three of these parameters.) In 1955, *National Safety News* wrote that "the instrument makes it possible to preserve records for evaluation of aircraft performance, stresses encountered in flight, . . . severe atmospheric disturbances, landing shocks, and other flight conditions."[38]

More so than any previous mechanical flight recorder, Ryan's invention was built for data indestructibility. In a paper presented in April 1955 at the American Society of Mechanical Engineers Diamond Jubilee Spring Meeting in Baltimore, Ryan described a few of the instrument's unique design features: "[The flight recorder] is spherical in shape to maintain the greatest rigidity against impacts due to aircraft crash. Between an internal and external sphere is a 1 in.-thickness of perlite insulation which is sufficient to prevent melting of the aluminum foil inside the recorder after 1/2-hr exposure to an intense heat of a fuel fire."[39] Elaborating further, he detailed the recorder's extraordinary capacity to withstand tremendous forces and to remain functional in hostile environments:

> The instrument functions over the range of ambient temperatures from −30 to +50 C and is not affected by extreme exposure of −65 C to +70 C. The recording medium . . . is not subject to deterioration or distortion. . . . The unit functions properly, and is not adversely affected, when exposed to extreme conditions of humidity as normally specified. The recording medium remains intact so that the intelligence can be analyzed when the instrument is subjected to a 100-g impact shock, and when exposed to flames at 1100 C for a period of 30 mins. It also is capable of remaining permanent and reproducible after exposure to a 36-hr period of immersion in sea water.[40]

Ryan could wax authoritative about his invention's rather astounding tolerances and endurances because it had been subjected to a battery of scientific tests, both in the laboratory and in the field. The tests were of three kinds: calibration, environment, and flight. The calibration tests microscopically measured "the displacement and the true position" of the recording styluses as a function of time.[41] The environment tests, conducted by the Civil Aeronautics Administration Technical Development and Evaluation Center, investigated the device's ability to endure extremes of temperature, vibration, and humidity. "The instrument survived a crash of 97 g, immediately followed by insertion in a large gasoline-fired torch for ½ hr. The record, although annealed, remained permanent and reproducible after these tests," according to Ryan.[42] The flight tests, in contrast to the environment tests, examined the recorder's accuracy and dependability under "normal" conditions of operation (that is, during routine flights). By mid-July 1954, device prototypes had logged more than 700 flying hours aboard commercial aircraft, as well as another 1,200 or so aboard noncommercial aircraft, with no sign of malfunction.

In the United States, the commercial, military, and governmental desire for a technology capable of automatically generating "an absolute and permanent record" of an unfolding aviation catastrophe dates back to at least the 1930s. Essentially the same inclination, the same *forensic impulse*, motivated the invention, a century earlier in England, of Charles Babbage's self-registering apparatus for railroad trains. In each case, mediated tracings were interpreted as material traces, machine "intelligence" (Ryan's term) as "mechanical evidence" (Babbage's term). It was hoped and believed that such evidence, by revealing the physical causes of accidents of accelerated mobility, could be used to reengineer faulty modes of transportation for greater safety and reliability.

Inoculation and the Indestructible Machine

In his paper's conclusion, Ryan tersely enumerated his instrument's purposes and capabilities. First item: "Provide a reliable means for analyzing and studying air failures." Final item: "Make all flying safer by the accumulation of this measured knowledge."[43] The General Mills Ryan Flight Recorder rearticulated the imagination of accident forensics in the context of postwar America. Even as the device's very existence reflected the perils of speed and the problem of the accident ("air failures"), its intentional design pointed to a future without slippage, a more secure tomorrow in which fast-moving bodies enjoyed maximal protection against the possibility of calamity ("flying safer"). Thus, Ryan's flight recorder, like all forensic media, materialized both a certain technological fear and a certain technological faith. According to this faith, which professed the remedial power of "measured knowledge," one disaster need not beget another; the cycle of disastrous repetition need never begin. "An attempt is made to insure that an air failure need only occur one [sic] to prevent its reoccurrence," Ryan proclaimed. "As Lord Kelvin so aptly pointed out, 'To measure is to know!'"[44]

The notion that "failure need only occur once to prevent its reoccurrence," that an accident learned from is a future accident frustrated, a reiteration interrupted, adheres to the logic of what Roland Barthes calls "inoculation." For Barthes, inoculation is a mythologizing stratagem that "consists in admitting the accidental evil . . . the better to conceal its principal evil. One immunizes the contents of the collective imagination by means of a small inoculation of acknowledged evil; one thus protects it against the risk of a generalized subversion."[45] In other words, when something normatively bad happens—something that upsets, offends, or threatens dominant social values, ideals, attitudes, or meanings—the stability and smooth continuity of the cultural order are most efficiently achieved when the existence of the bad something is conceded and confessed rather than ignored or refused. The badness must be conceded and confessed in a very particular way, however. The cultural order itself, as a structured and structuring totality, must not be prodded or probed, must not be made to submit to searching inquiry and analysis, much less condemned or attacked outright. Its fundamental deficiencies (inadequacies, inequities, illogicalities) must not be called into question, as doing so risks precipitating a mass revolt. Instead, a "bad apple" must be singled out, individualized, branded and isolated, constituted as an exception. This "small inoculation" thereby ensures the survival of the ideological system, the preservation and perpetuation of the status quo.

In conformity with the gospel of "learning through failure," forensic mythology admits "the accidental evil" of the technological accident.[46] But it does so cannily, strategically. Recall how the flight recorder (of whichever kind) was discursively constructed and contextualized in the various aforementioned newspapers, magazines, trade journals, and technical reports. There, and wherever it appears, the flight recorder is a contrivance whose purposeful objecthood tacitly acknowledges the "evil" of the plane crash. Its artifactuality effectively attests the distressing reality of human error and mechanical failure. Hence, every account, every figuration, every discussion, every last mention of the flight recorder necessarily evokes, if only dimly, obliquely, if only *unconsciously*, a technological lifeworld in which disaster strikes, things fall apart, shit happens. Yet such evocations do no real or lasting harm. They elicit no substantive critique of technological approaches or objectives; they effect no "generalized subversion" of technological rules and relations. On the contrary, they swiftly administer an ideological inoculation: the disease of accident is delivered, but only in a small, salutary dose; "the collective imagination" is exposed to danger and trauma, but not too much danger and trauma; the system is shocked, but in a way that renders it more shock-resistant thereafter. A bad, technologically treacherous world is conjured, to be sure, but only and always in terms of a technical problem to be solved, a defect or malfunction to be fixed. Or, rather, that world is conjured only and always in terms of a technical problem that *can* and *will* be solved, a defect or malfunction that *can* and *will* be fixed, with the aid of knowledge acquired through a forensic medium such as the flight recorder. In the end, the "principal evil" that forensic mythology "conceals"—conceals precisely (and paradoxically) in the act of *singling it out*, understood in the double sense of identifying it and excepting it, of nominating it and anomalizing it—is the accidentality of the accident: its irreducibility, its ineliminability, and, despite Ryan's dream of an instrument that would assist in preventing failure's recurrence, its ceaseless iterability.

Besides rearticulating aspects of forensic mythology, the General Mills Ryan Flight Recorder rehearsed the trope of the indestructible machine, the invulnerable technology. In the West, this trope extends at least as far back as 1482, when Leonardo da Vinci, in a letter to Ludovico il Moro, Duke of Milan, wrote of having prepared plans for bridges that are "secure and indestructible by fire and battle," for "vessels which will resist the fire of the largest cannon, and powder and smoke," and for "covered cars, safe and unassailable, which will enter among the enemy with their artillery, and there is no company of men at arms so great that they will not break it."[47] Beyond Leonardo and the Italian Renaissance, the notional nexus between machinic unassailability and military vessels/vehicles expanded and intensified with developments in battlefield technology during the twentieth century. From First World War tanks such as the British Mark I ("Big Willie," "Mother") to Second World War aircraft such as the American B-17 "Flying Fortress" to post-Cold War submarines such as the Russian *Kursk*, the idea of the mobile, heavily armored, virtually invincible war machine has repeatedly exercised the modern military-industrial imagination.[48]

Apart from its martial incarnations, the indestructible machine has been a popular topos in news and entertainment media throughout the twentieth century. The RMS *Titanic*, prior to its fatal collision with an iceberg on 14 April 1912, was widely hailed as "unsinkable," a mechanical marvel of unparalleled size, strength, and security. The *Titanic* tragedy, however, did not destroy the currency of the trope of the invulnerable technology, as even a cursory inspection of science-fiction themes and iconography confirms. Indeed, across the entire spectrum of science-fiction forms (novels, comic books, movies, radio programs, television shows, videogames), the android, "replicant," or spaceship that proves impervious to all manner of attack—thanks to its impenetrable armor, electromagnetic force field, or other means of fortified protection—has been a hallmark of the genre. Probably the most famous, and certainly the most spectacular, examples have come from Hollywood films: Gort, the towering robot in *The Day the Earth Stood Still* (1951), and the Terminator, the unstoppable cyborg from the 1984 blockbuster (and its several sequels) of the same name.

The discursive history of the phonograph offers another kind of example.

During the period of its emergence, the phonograph was frequently invested with fantasies of preservation and permanence. As Jonathan Sterne observes, "From the moment of its public introduction, sound recording was understood to have great possibilities as an archival medium. Its potential to preserve sound indefinitely into the future was immediately grasped by users and publicists alike."[49] In an 1878 article for the *North American Review*, the phonograph inventor Thomas Edison declared, "Repeated experiments have proved that the [record's] indentations possess wonderful enduring power, even when the reproduction has been effected by the comparatively rigid plate used for their production."[50] Pushing the promise of medium robustness to its logical limit, one Albany, New York-based company went so far as to name itself the Indestructible Phonographic Record Company. According to Sterne, "The company's advertisements emphasized the durability of the cylinders and played on the idea of indestructibility with pictures of a child putting a stick of dynamite into a cylinder or polar bears rolling around on Arctic ice with one in a cylinder, although the process or the material composition of the cylinders was never explained."[51] Here we are invited to consider a recording medium that can withstand even a violent explosion, that can survive even the most frigid immersion (in the ad, two polar bears are in the process of rolling a third bear, encased in an oversized phonographic cylinder, off the edge of an ice sheet and into what is presumably the Arctic Ocean). Though aimed at a very different audience, these promotional images from 1908 strikingly anticipate Ryan's contention, advanced forty-seven years later, that his flight recorder's medium is functional after being subjected to "a 100-g impact shock," "flames at 1100 C for a period of 30 mins," and "a 36-hr period of immersion in sea water."

Whereas the Indestructible Phonographic Record Company's advertisements were to be taken with a grain of salt, the claims of indestructibility made for Ryan's device, whether proffered in technical papers or publicized in the press, were to be taken as demonstrated fact. Amusing and obviously hyperbolic, the former would not have convinced a reasonable consumer that the company's cylinders were really dynamite-proof, frost-proof, or seawater-proof. Clearly, the ads' point—their pitch—was that, compared to other commercially available sound-recording media (made from inferior materials: wax, tinfoil, paraffin), Indestructible's cylinders (made from celluloid) were tougher, sturdier, and lasted longer. Something like "artistic license" allowed the company to associate its product now with a child lighting a stick of dynamite, now with a trio of polar bears romping on an Arctic ice sheet. General Mills and James Ryan, on the other hand, were neither in need of, nor entitled to, any such license when describing their instrument's properties and capacities. Their descriptions, precise and propositional, were to be received with due seriousness, issuing as they did from authoritative social actors (a major corporation with research-and-development ties to the U.S. military, a senior corporate engineer and university professor) and based as they were on scientific evidence and experiment.

Throughout their history, recording media, from clay tablets to phonographic cylinders to compact discs to the latest digital-storage software, have been imagined as preservative and/or permanent.[52] Preservation and permanence are interrelated but differently inflected concepts. While each is concerned with archivability, "preservation" accents the storage medium's informational indelibility, its resistance to decay, whereas "permanence" accents the storage medium's physical durability, its resistance to damage. The General Mills Ryan Flight Recorder simultaneously radicalized, recontextualized, and redirected these ideas and imaginings. Its archival medium held up under extremely adverse conditions (high-speed impact, blazing fire, freezing cold, seawater immersion)—but more than that, it was literally built for them. Equipped for every conceivable aviation catastrophe, its mission was to go to hell and back, as it were, and to faithfully report on what it had "witnessed" while on assignment. The issue—both the media-technical problem and the media-cultural problematic—was not about "ordinary wear and tear," not about longevity or lastingness in relation to "normal use." Rather, it was about the ability of the device's hardware to endure the severest of trials, the most difficult of ordeals, the so-called worst-case scenario, as well as the ability of its software to remain not only "permanent and reproducible," as Ryan put it, but also free from "deterioration or distortion" in the aftermath of those trials and ordeals. In the domains of popular culture and commercial entertainment, the fantasy of recording-media indestructibility was linked, then as now, to the ideal of playback "fidelity" and to the promise of infinite repeatability.[53] In the domains of science, industry, and government, by contrast, the fantasy of flight-recorder indestructibility—a fantasy that assumed programmatic proportions in the mid-1950s—was linked, and continues to be linked, to the possibility of forensic discovery through mechanical reliability and reproducibility.

By the end of the decade, General Mills had sold the design for its flight recorder to Lockheed Corporation, which "produced it as Model 109-C until 1969."[54] Meanwhile, the Civil Aeronautics Board issued another round of regulations in 1957, this time mandating that all aircraft weighing more than 12,500 pounds and carrying passengers above 25,000 feet be equipped with a crash-protected, five-parameter, foil-oscillographic flight recorder by 1 July 1958. The regulations demanded compliance with Technical Standards Order

TSO-C51, which specified the instrument's recording parameters (time, altitude, airspeed, compass heading, vertical acceleration), sampling interval, range of accuracy, and survivability thresholds.[55] Even more than in the previous decade, the issuance of new CAB regulations invited market competition and commercial innovation, and within a few years several companies besides Lockheed were manufacturing TSO-compliant flight recorders, including Sundstrand, Waste King, Allied Signal, Fairchild Aviation, and Minneapolis-Honeywell.[56]

In August 1958, President Dwight D. Eisenhower signed into law the Federal Aviation Act, transferring from the Civil Aeronautics Board to the newly created Federal Aviation Administration (FAA) the responsibility for issuing safety rules and regulations. Issued by the FAA in 1966, Technical Standards Order TSO-C51a, in addition to requiring that an airplane's flight recorder be located as far aft as practicable, appreciably increased two of the performance standards specified in TSO-C51—impact shock (from 100 Gs to 1,000 Gs for five milliseconds) and seawater immersion (from thirty-six hours to thirty days)—and added three more for good measure: static crush (5,000 pounds for five minutes on each axis), impact penetration (500 pounds dropped from ten feet with a quarter-inch-diameter contact point), and corrosive fluids (immersion in oil, fuel, and other aircraft fluids for a period of twenty-four hours).[57]

The Military/Maritime Phonograph

Reporting on military affairs in England in 1888, a foreign correspondent for the *New York Times* recounted a conversation he had had with a British officer concerning the naval and military possibilities of Edison's "talking machine":

> A distinguished officer said to me, after we had inspected Edison's phonograph . . . , "I wonder can that be adapted to naval and military use?" Subsequent conversation with him developed these ideas. There is hardly a campaign of which we have a complete record; there is hardly an accident at sea of which we have ever heard, or that did not give rise to the question of the exact wording and precise emphasis on the wording of an order. . . . Now, if there were a portable form of the phonograph there could be a record of the orders given, whether in case of an action ashore or an accident afloat. Perhaps Mr. Edison will condescend to give the subject his attention. . . . But in applying the phonograph to military and naval purposes it does seem to be necessary that there should be found some more durable, even if less sensitive, material than the wax cylinder as the receiver of the vibrations.[58]

Five years later, the *Phonogram*, a short-lived periodical focusing on the phonograph's business applications, ran an article titled "The Phonograph at Sea," which speculated as to the medium's maritime utility:

> We know that if several records of the condition of the disabled vessel had been made by some one on board capable of accomplishing this—before the danger became imminent, and these waxen cylinders had been placed in tin cases constructed so as to be corked like a bottle or jar, and sealed with rosin or any other preparation that would render them water-proof, these cases floating along in the path followed by nearly all the steamers crossing the Atlantic, would by this time have been discovered, and we should now be in possession of some meagre details of the terrible catastrophe; or at any rate might gather from the records of those looking their doom in the face, facts going to suggest alterations in the equipment or arrangement of other transatlantic liners that would save them from similar casualty.[59]

The Phonograph's Strange Audience.

Illustration accompanying "The Phonograph at Sea," in *Phonogram* (March–April 1893). Courtesy of Smithsonian Libraries.

New media and communications technologies do not spring forth fully formed like Athena from the forehead of Zeus. Their material configurations are no more naturally preordained than their social functions are technically predetermined or their cultural meanings universally predefined. Their uses, their "purposes," are neither handed down from on high nor chiseled into granite. As Geoffrey Pingree and Lisa Gitelman observe, "When new media emerge in a society, their place is at first ill defined, and their ultimate meanings or functions are shaped over time by that society's existing habits of media use (which, of course, derive from experience with other, established media), by shared desires for new uses, and by the slow process of adaptation between the two. The 'crisis' of a new medium will be resolved when the perceptions of the medium, as well as its practical uses, are somehow adapted to existing categories of public understanding about what the medium does for whom and why."[60]

Written and circulated within the first two decades of the phonograph's invention, the *Times* and *Phonogram* articles indicate the extent to which the medium's practical applications and social contexts of use were still open to interpretation, subject to revision, and amenable to imaginative leaps. They suggest that the phonograph's "place" was not yet fixed, either in the popular consciousness or in the commercial marketplace. They imply that "proper" phonographic attitudes, protocols, functions, and conditions were still being hammered out, even as the forces of normalization (industrial standardization, institutional regulation) were working to hem them in. Simply put, the phonograph as a cultural technology was then in "crisis."[61]

Edison initially conceived the phonograph as a machine for recording telephone messages. In his aforementioned 1878 article (which actually was ghostwritten by his business associate, Edward H. Johnson), the inventor expanded his initial conception to encompass a number of other potential uses for sound-reproduction technology, such as letter writing, dictation, books for the blind or infirm, elocutionary instruction, the reproduction of music, the "family record" (a phonographic album of family members' voices, sayings, and dying words), music boxes and toys, clocks that announce the hour of the day, commercial advertising, and the preservation of political speeches.[62] As David Morton notes, "What Edison had first conceived narrowly as a telephone recorder for business he now predicted would become a more general-purpose enhancement or even replacement for many kinds of oral and written communication for business and personal purposes."[63]

The ideas presented in the *Times* and *Phonogram* at once resonated with and reached beyond the hodgepodge of conceits, conjectures, and prophecies that Edison had set forth in the pages of the *North American Review* some years prior. In the first instance, they echoed what Edison called "the almost universal applicability of the [phonograph's] foundation principle, namely, the gathering up and retaining of sounds hitherto fugitive, and their reproduction at will."[64] The implicit reasoning was simple enough: war theaters and ship decks each resound with voices and myriad other audible emanations, and these, like all "sounds hitherto fugitive," were available—indeed, why not?—for phonographic collection, retention, and "reproduction at will." Doubtless, for the *Times* and *Phonogram*, the medium's capacity to snatch and hold sonic vibrations was self-evident, as it was for Edison, who insisted that the phonograph made possible "the captivity of all manner of sound-waves."[65] In this sense, the articles endorsed the phonograph's "foundation principle."

Edison was fond of claiming that his invention was "practically perfected in so far as the faithful reproduction of sound is concerned," that it provided "an unimpeachable record," and that it functioned as a sort of "conscientious and infallible scribe."[66] Similarly, the *Times* promised that the phonograph, were it to be brought either into battle or aboard a seagoing vessel, would provide "a complete record" of important oral communications such as military commands. Its mechanical ear, which "heard" by means of a mouthpiece, would pick up and preserve the diction and inflections of human speech, the "exact wording and precise emphasis" of spoken utterances. *Phonogram* contended much the same, equating a phonographic "record of those looking their doom in the face" with an oral registry of "details" and "facts." Such truth-claims were in sync with Edison's contention that his talking machine not only accurately reproduced a wide range of speech forms, including "interjections, explanations, emphasis, exclamations," but also occasionally improved their articulacy and intelligibility:

> The writer has at various times during the past weeks reproduced [sound] waves with such degree of accuracy in each and every detail as to enable his assistants to read, without the loss of a word, one or more columns of a newspaper article unfamiliar to them, and which were spoken into the apparatus when they were not present. . . . Indeed, the articulation of some individuals has been very perceptibly improved by passage through the phonograph, the original utterance being mutilated by imperfection of lip and mouth formation, and these mutilations eliminated or corrected by the mechanism of the phonograph.[67]

Yet, as much as the *Times* and *Phonogram* reinforced normative epistemologies of the phonograph—epistemologies authorized by Edison from the very start—they also pushed the envelope by recommending for the medium an alternative matrix of rationales, modalities, and spheres of application. In the last decades of the nineteenth century, Edison's phonograph and its competitor, Alexander Graham Bell and Charles Sumner Tainter's "graphophone," were manufactured and marketed primarily as machines for dictation and record-keeping and secondarily (albeit increasingly with each passing year) as machines for popular amusement. Phonographic dictation was done in business offices and, for some well-to-do consumers, in private homes; "nickel-in-the-slot" phonographic entertainment was had in barrooms and, as David Nasaw states, "on fair midways, in train stations, in hotel lobbies, and at summer resorts."[68] The *Times* and *Phonogram*, however, asked the phonograph to undertake obligations, execute functions, and operate in circumstances of entirely different kinds. In fact, what they proposed for the medium could scarcely have been less like popular amusement. It differed greatly from office dictation, too. As if testing Edison's proposition of "universal applicability," the articles advocated the extension of the phonograph's reach—practical and political—into realms untried, uncertain, and theretofore unimagined.

The call to render the phonograph serviceable to military/maritime interests and institutions is historically and culturally significant. Exemplifying what Edison had described as "the imaginative work of pointing and commenting upon the possible," the *Times* and *Phonogram* articles represented early attempts to envision and publicly articulate how sound-reproduction technology might be mobilized to meet the many and diverse security needs of a modern industrialized nation and society.[69] In a military setting, the phonograph could perhaps be counted on to keep an objective log of orally communicated "orders given" by commanding officers, thereby eliminating after-the-fact ambiguity and ensuring military-historical accuracy (who said what, when, and to whom). In a maritime setting, the phonograph could perhaps be employed by the captain or a crewman to chronicle the occurrence of an unfolding tragedy, and that chronicle could, in turn, be employed by naval architects and shipbuilders "to suggest alterations in the equipment or arrangement of other transatlantic liners that would save them from similar casualty." Each of these imaginings, so remarkable for its day, embroiled the new recording apparatus in the logics and politics of risk, safety, and protection: the one by installing it at the primal scene of national security, the battlefield; the other by instrumentalizing it for the "deterrence" of nautical disasters. Liberated from its pedestrian applications in the worlds of white-collar labor (office dictation) and working-class leisure (popular amusement), the phonograph here becomes something far more extraordinary and, at the same time, far more indispensable. It becomes a means of defending the nation, of guarding against danger, of protecting bodies from harm, of saving lives. It becomes a technology for the production of historical veracity and for the prevention of future calamity. It becomes, in short, a forensic medium.

The *New York Times* and *Phonogram* articles are especially noteworthy, then, because they show that the history of the idea of using sound-reproduction technology to achieve both retrospective certainty (truth) and prospective security (progress) vis-à-vis modern transportation mishaps—a dual achievement predicated on the medium's supposed power to capture and preserve the voices of catastrophe's imminent victims, the utterances of the doomed—can be traced back to the first decades of Edisonian sound recording, even if the practical and institutional realization of that idea had to wait until the advent of the cockpit voice recorder in the mid-1950s.

Also noteworthy is the suggestion that the phonograph, in order to be usable in a war zone or on the high seas, would require some measure of technical modification, some degree of redesign. Specifically, it would need to be *ruggedized* (to use an anachronistic term advisedly).[70] When Edison asserted in 1878 that the phonographic record "possessed wonderful enduring power," he was referring to the wax cylinder's ability to continue to function properly—to reproduce sounds "faithfully"—after repeated plays. Quite evidently, he did not have in mind operating conditions as harsh or inhospitable as those specified in the *Times* and *Phonogram*. Regarding such conditions, the *Times* was concerned about the robustness of the recording medium, concluding that "it does seem to be necessary that there should be found some more durable, even if less sensitive, material than the wax cylinder as the receiver of the vibrations." *Phonogram*, for its part, was concerned about the durability of the medium's container as opposed to that of the medium itself, counseling that the "waxen cylinders [be] placed in tin cases constructed so as to be corked like a bottle or jar, and sealed with rosin or any other preparation that would render them water-proof."

Here we glimpse perhaps the earliest public expression of what would come to be regarded as a mandatory requirement for all transportation recorders, including the cockpit voice recorder: data survivability. (Interestingly, Charles Babbage made no mention in his autobiography of having so much as considered survivability criteria or the question of crash protection when building and testing his self-registering apparatus in 1839.) Since both the veracity of the past and the security of the future were thought to be at stake, all suitable precautions had to be taken so that the acoustic traces, the phonographic "indentations," would emerge from the ordeal essentially unscathed. Indeed, according to what might be called the *survivability imperative*, the destructive violence of the accident had to be planned for ahead of time, deliberately accommodated in the apparatus's engineering and design. The consequences of catastrophic failure had to be preemptively neutralized, either by making the medium ruggeder (per the *Times*) or by ensuring the sturdiness and imperviousness of its outer shell (per *Phonogram*). Thus, not only do these articles anticipate the problem of recording-media "crashworthiness"; they also anticipate its twin solutions: strengthen the software or harden the hardware; reinforce the information's carrier or its carapace; immunize the interior or "armorize" the exterior.

The notion that a firsthand account of a maritime accident could be preserved by hermetically encasing it in something "like a bottle or jar"—which hermetic encasement (a whiff here of the impermeable seal's ancient occultism), having been tossed into the ocean, might later be found "floating along in the path" of a steamship—was nothing new. On the contrary, the act to which *Phonogram* alluded—namely, sending a handwritten message in a bottle—has been performed for thousands of years. As the *Bulletin of the International Oceanographic Foundation* explains in an article titled "Neptune's Sea-Mail Service," "Seaborne bottles bearing notes have often been associated with shipwrecked seamen cast ashore on some uninhabited island, launching a last empty bottle in the hopes of a rescue. Drift bottles have been used for centuries, however, to shed light on nautical mysteries, to convey secret messages, or to seek contact with an anonymous pen pal in a faraway country."[71]

For *Phonogram*, the invention of the phonograph offered an opportunity to modernize—in fact, to mechanize—the practice of sending a distress message in a bottle. Previously, message-in-a-bottle recipients had to rely on fragments of information scrawled on scraps of paper for an indication of what caused the ship to sink. Hastily written, truncated or overly abbreviated, and sometimes barely legible, this sort of scribble was of limited value to those who wished to reconstruct the story of the accident. On the other hand, a message recorded at the speed of speech and "bottle[d] up for posterity," to use Edison's fitting phrase, would be capable of carrying considerably more information than a handwritten note, and would be clearer and more comprehensible to boot.[72] Reliably transcribed by a sonic stenographer, securely transported in a watertight container, such a message promised "to shed light on nautical mysteries" in a manner until then impossible.

The ARL Flight Memory Recorder, circa 1958. Courtesy of Defence Science and Technology Organisation, Australian Department of Defence.

David Warren's Device

In April 1954, David Warren, an Australian chemist specializing in aircraft fuels, prepared a technical memorandum for Aeronautical Research Laboratories in Melbourne, Australia. The four-page report, titled *A Device for Assisting Investigation into Aircraft Accidents*, did not deal with issues pertaining to aviation chemistry, as might be expected given its author's area of specialization. Instead, it succinctly stated the case for an onboard technology that would continuously record the voices of a pilot and a copilot as they operated an aircraft. Should disaster strike, the record of those voices would be of "inestimable value" to accident investigators, Warren reasoned, because it would contain otherwise unobtainable evidence as to the cause of the crash:

> It may be assumed that in almost all accidents the pilot receives some pre-indication either by sight, feel of controls, automatic alarm or instrument reading. In most cases this would evoke a complaint of difficulty or a shout of warning to attract the attention of the co-pilot. Unless radio contact is actually in progress there is often not time to get any information through before the crash. . . .

> In the case of fire [there] would almost certainly [be] a shout from the first crew-member to detect it, followed by verbal instructions. Careless control or error-of-judgment (as is often suspected in landing and take-off accidents) would probably elicit criticism, suggestion or warning from the co-pilot. An unexpected fuel-tank explosion would be recorded as an interruption of normal conversation by the first part of the explosion noise followed by immediate cut-out.[73]

Warren recommended that a magnetized steel wire be used as the recording medium. The wire would run across an erasing head and a recording head in a continuous loop, with new data replacing old every two or so minutes. In the event of an accident, the looping mechanism would switch off automatically, so as to "provide a permanent 'memory' of the conversation in the control cabin" during the final moments of flight.[74] The device itself would be small (occupying less than 0.1 cubic feet of space), lightweight (less than five pounds), and inexpensive (sixty dollars per dozen). For partial power, it could be hooked into the airplane's radio receiver, and it would be easy to maintain, needing only "an occasional check that it was in working order."[75] What is more, the steel-wire medium "would not be greatly harmed by impact nor by suffering moderate heating," though Warren conceded that "the extent of the latter would need to be checked by experiment."[76] To ensure its intact recovery after a crash, the device could be installed in the tail section of the plane, where it would likely sustain the least amount of damage. A method might even be devised to automatically eject it at some point prior to impact, thereby severing its fate from the presumably worse one awaiting the aircraft that housed it. "If the containing box were reasonably robust," Warren mused, "no parachute would be required, as only the wire need be salvaged. An attached marker streamer, however, would greatly help in finding the unit."[77]

Warren foresaw only one potential problem in all of this: cockpit voice recording, as it would come to be known, might have a negative "psychological effect" on the pilots.[78] Oral exchanges between pilot and copilot would no longer be private or effectively privileged. Quite the opposite: they would be subject to external scrutiny, exposed to the surveillance of anonymous others. Making a virtue of necessity, the resourceful inventor found a remedy for this hypothetical difficulty in the mechanism's technical limitations: "The possible objection by crew to having their conversation continually recorded is countered by the fact that the device has such a short memory. If no accident occurs, anything said during flight is obliterated during the time taken to taxi in."[79]

Warren was surely unaware that an idea very much like the one he was put-

picks up and preserves "complaints of difficulty," "shouts of warning," maybe even a "fuel-tank explosion." It is there and not there, like a fly on the wall. It is a bug.

Unlike the *Times* and *Phonogram* articles, Warren's report raised the specter of surveillance. It acknowledged that the flight crew might resent the installation of an officially mandated, automatically operated, inaccessibly located recording apparatus. It admitted that the pilots might resist the invasion of their privacy, the incursion of Big Brother into their workplace. Would the skulking presence of the device undermine morale? Would it arouse indignation? Would it diminish performance or productivity? Worst of all, would it serve to discourage inter-pilot communication, thereby leading to a decrease in operational safety, thereby leading to an increase in accidents—accidents whose elimination, ironically, it had been put there precisely to accomplish? By taking such questions into consideration, if only implicitly, the report forged discursive and programmatic links between aviation surveillance and audio surveillance.

The concept and practice of using flight recorders to invigilate airplane pilots preexisted the appearance of Warren's technical memorandum. Indeed, as we have seen, flight-recorder surveillance had been around since at least 1937, when the *New York Times* and *Science* each reported on the development of an instrument designed partly to deter dangerous flying by providing authorities with an onboard means of monitoring pilots' compliance with safety rules and regulations. This was monitoring through mechanical inscription, with its inking pens and squiggly lines and scrolling paper. Warren's memorandum, by contrast, proposed an onboard-monitoring instrument whose medium would be sonic and magnetic rather than graphic and mechanical, whose data would be verbal not visual, audible not tangible, reproductive of human speech not representative of nonhuman behavior (the plane's performance). Moreover, it suggested, contrary to the *Times* and *Science* articles, that flight-recorder surveillance might be a matter of some concern, even controversy. In 1937, surveillance could be promoted as one of the principal benefits of flight recorders, but in 1954 it was viewed as a potential problem. What, perhaps, accounts for this discrepancy?

The flight recorder of 1937 was an early flight-*data* recorder. Its task was to register the technical performance, the various motions and varying conditions, of an inanimate object, and to represent the results as curvilinear tracings on paper. Human agency does not immediately figure into the equation here. No intentional subjects, no interposing bodies. Just one machine monitoring the functioning of another. Automated operations, automated output. Flight recording is a fundamentally impersonal process. Warren's device con-

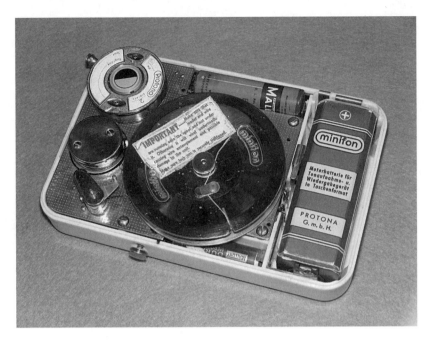

Minifon wire recorder, circa 1953.

figured a very different set of technical functions and cultural associations. Its task was to grab sounds from the air and save them to steel wire. Because it was a cockpit *voice* recorder, this task was necessarily tied to human subjectivities and bodies—the voice being traditionally construed as the supreme expression of human agency and intentionality, as that which stands metonymically for human identity. The introduction of intentional subjects and interposing bodies thus radically alters the equation. Using a mechanical instrument to monitor and measure the plane's performance is one thing; using an electromagnetic instrument—a sneaky little gadget—to "steal" and store the pilots' speech is quite another. When flight recording is machine on machine, when its operations and output are entirely automated, it seems to enjoy a certain amorality. But as soon as the human voice enters the scheme of things, the *drama* proper begins: flight recording is suddenly freighted with ethical questions, imbued with metaphysical connotations. In seizing hold of someone's voice—someone who has no real say in the matter—flight recording threatens an affront to dignity, a breach of privacy, an assault on liberty, even a theft of identity. What had been fundamentally impersonal becomes disconcertingly personal.

Pocket-Size Wire Recorder

PERHAPS one of the most sensational units to appear on the wire-recorder scene recently is a complete battery-operated recorder, 6¾ x 4⅜ x 1½ in. in size. It records, erases and plays back through a pair of lightweight earphones.

The entire recorder fits any average-size pocket, or it can be carried and operated in a fabric shoulder-type carrying case, as illustrated in photo A. Two types of sensitive miniature crystal microphones are available, as shown in photo D. One is a lapel variety and the other is a wrist-watch type worn by the operator in photo A, for making concealed recordings useful in detective work and for checking comments in crowds at shows and similar applications.

This Minifon recorder, made in Germany, is now available on the American market; it is powered with standard miniature A and B-batteries. The motor is driven by a Mallory mercury-cell-type battery pack that sells for $4.25. This provides 24-hour service. The A and B-batteries last for full shelf life. An a.c. power-supply unit also is available for operating the motor from 110-120 volt a.c. lines. Photos B and C are internal and external views of the recording and playback unit. Recording wire is available in spools providing ¼ to 2½ hours of continuous operation.

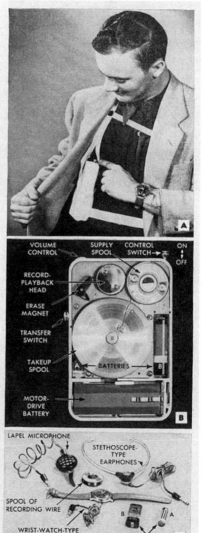

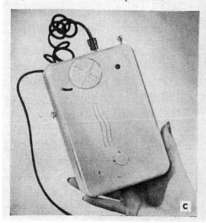

Article on the Minifon recorder. Source: *Popular Mechanics* (August 1953).

Minifon P55 marketing brochure, circa 1955.

In Warren's report, then, the recording of cockpit conversations was viewed as a potential problem, in part, because of the moral and metaphysical baggage borne by the human voice. Insofar as it problematized the involuntary and indiscriminate recording of human speech, the report participated in a long line of thinking about the ethical and political implications of audio surveillance. It also discursively articulated the emergent practice of cockpit voice recording to older practices of electrical eavesdropping such as telegraphic and telephonic wiretapping (the former was employed by both the North and the South during the American Civil War), as well as to the practice of "making of secret or surveillance records," which, as David Morton observes, dates back to "the early days of the phonograph, although it required a considerable amount of naiveté on the part of others."[80]

Warren's device's relationship to surveillance runs deeper still. Warren, who as a young boy built and sold crystal radio receivers, modeled his invention on the Minifon, a miniature, battery-operated wire recorder developed in Germany in the early 1950s. Initially marketed as a portable dictation device, the Minifon soon found other, less "legitimate" applications, including spying and surreptitious recording. Morton explains: "The idea of offering a pocket dictating machine was novel, since dictation had previously been done in the office. However, it was thought that people like salesmen could take the machine 'on the road' with them. Once on the market, the Minifon's promoters discovered that many people took advantage of the recorder's small size to make secret recordings to be used as evidence, as in court."[81] By the time Warren wrote his technical memorandum—a memorandum in which he duly noted "the possible objection by crew to having their conversation continually recorded"— miniature wire recording was already allied with audio surveillance through the "illegitimate" practices of some of its users.

In a broader sense, Warren's device—and the same must be said for Ryan's— mirrored the Cold War era's massive, multifarious investment (economic, psychological, bureaucratic, ideological) in the protective possibilities of high-tech monitoring and tracking, of electronic data gathering and processing, of communications interception and decryption. In harmony with the practices of the emergent national-security state, and with the ascendant ethos of information feedback and control, these postwar flight recorders, in their own particular ways and contexts, rationalized techniques of surveillance and realized a strategy of deterrence.

A Device for Assisting Investigation into Aircraft Accidents circulated through the Australian aviation industry in 1954. It also went out to Australia's Department of Civil Aviation and the Royal Australian Air Force. Warren received no replies. He surmised "that the lack of interest might stem from the fact that Australia hadn't had a major air accident in three years. At the time there was a general feeling that Australia had quite safe skies."[82] Warren now resolved to cast a wider net, sending his report to all the world's major airlines and to the Federal Aviation Administration in the United States and the Air Registration Board in the United Kingdom. Still no replies.

In 1958, with the support of his superintendent, Tom Keeble, and the assistance of Titch Mirfield, an instrument engineer, Warren built a prototype, using Minifon components, for the purpose of demonstration. More advanced than the device described in his original memorandum, the ARL Flight Memory Recorder, as it was dubbed (the initials standing for Aeronautical Research Laboratories), could record four hours, rather than two minutes, of cockpit conversation. In addition, it could store readings from eight flight instruments, making it not just the first cockpit voice recorder but the first combination cockpit voice and flight-data recorder.[83]

The official responses to Warren and Mirfield's prototype were less than enthusiastic, this despite the fact that it had performed admirably during its flight tests. "We finished developing this to this stage and found that people weren't interested," Warren stated in an interview for Australian television, broadcast in November 2002. "We wrote to the Department of Civil Aviation and the letter we got back said, 'Dr Warren's instrument has little immediate, direct use in civil aviation.'"[84] The Royal Australian Air Force concluded that "such a device is not required," since in all probability it "would yield more expletives than explanations."[85] The Australian Federation of Air Pilots complained that "it would be like having a spy on board—no crew would take off with Big Brother listening."[86] (Warren's "psychological effect" remark proved prescient.) And the Aeronautical Research Council decided that, in view of the alleged difficulties involved, "no further action should be taken."[87]

Robert Hardingham, Secretary of the U.K. Air Registration Board, responded a good deal more favorably to Warren's device. Having taken curious notice of it during a visit to Aeronautical Research Laboratories in 1958, Hardingham arranged for the device to be brought to England for a formal demonstration. (Do we detect here the ghostly trace/transmission of Marconi, who, some sixty years previous, likewise brought his black box to England for a formal demonstration?) According Macarthur Job, author of *Air Disaster*, the demonstration was a success by any standard: "The BBC reported on it, aircraft manufacturers were supportive, aviation authorities considered making flight recorders mandatory, and the firm of S. Davall & Son bought production rights, subsequently developing a crash recorder that won a major share of the British market."[88] The design for S. Davall & Son's "Red Egg" recorder, so nicknamed for its red-painted exterior, was modeled, not on Warren and Mirfield's original demonstration unit, but on its more sophisticated successor, the Pre-production Prototype ARL Flight Memory Recorder, developed in 1962 by Warren and Aeronautical Research Laboratories associates Lane Sear, Ken Fraser, and Walter Boswell.[89]

In the United States, meanwhile, the FAA, in response to Civil Aeronautics Board recommendations, conducted a study in 1960 to determine the feasibility of recording flight-crew conversations for the purpose of accident investigation.[90] Two years later, the FAA carried out tests on a new model of magnetic-tape flight recorder manufactured by California-based United Data Control—the same company, as it happens, that had been commissioned two years earlier to develop a similar device for the Australian market (much to the chagrin of Warren and his colleagues). Hopes ran high for the new machine. The *New York Times* quoted an unnamed FAA official in April 1962: "The flight [data] recorder can tell us at the most what did not cause an accident. It is not likely to reveal the cause itself—as we hope a cockpit voice recorder will."[91] In his book on the Federal Aviation Administration, Robert Burkhardt writes that the agency "in 1965 ordered the installation and use of cockpit voice recorders in large transport airplanes operated by air carriers and commercial operators."[92]

Artifacts of Forensic Desire

In November 1943, *Life* magazine ran a story in its "Science" section about the use of a recently developed "magnetic wire recorder" for military observation flights. Just as the Brown Instrument potentiometer and the Vultee Radio-Recorder promised to make the tracking and transcribing of flight-instrument readings easier, more efficient, and less prone to error, so the wire recorder promised to increase the ease, efficiency, and accuracy of aerial reconnaissance. All three wartime aviation technologies promised improvements in record-keeping through automation. However, in the case of the wire recorder, the means of improvement involved sonic reproduction, not mechanical inscription: "Instead of making notes in a pad strapped to their right knee they speak their observations into a microphone which fits in the palm of the hand like a stop watch. Their words are magnetically recorded on a thin steel wire uncoiling between doughnut-size spools mounted in a compact case. The resultant spool of magnetized wire when played back into sound provides a fresh, on-the-spot account, far better than notebook reports."[93] The article goes on to say that the wire recorder might also find application *on* the battlefield, not just above it: "Highly portable, the recorder could be used for on-the-spot transcriptions of tank, artillery and naval actions. The War Department is already encouraging war correspondents to use it for dictating battle descriptions for subsequent broadcasts or transcription into written accounts."[94]

The wire recorder described in the pages of *Life* in 1943 constitutes a "missing link" of sorts between the late-nineteenth-century military/maritime phonograph and the mid-twentieth-century ARL Flight Memory Recorder. On the one hand, it rehearsed the decades-old dream of using sound-reproduction technology as something like a war-zone Dictaphone; on the other hand, it introduced magnetic-wire recording into the aircraft control cabin eleven years before David Warren wrote his technical memorandum. Also, interestingly, it connects to the early history of the flight-data recorder, insofar as its aeronautical applications paralleled the prescribed functions of the Brown Instrument and Vultee recorders.

Contrary to the moment-of-eureka mythology informing most accounts of the invention of the cockpit voice recorder, Warren's report was not the first printed source to imagine that sound-reproduction technology might be adapted and deployed for transportation-accident investigation (the *Times* and *Phonogram* beat it to the punch). Neither was his device the first wire recorder to be used in the cockpit of an airplane (as the *Life* article demonstrates). Wherein, then, lies the historical and cultural import of Warren's invention? While we are at it, let us also inquire as to the import of James Ryan's invention, that other postwar flight recorder to which claims of historical primacy have been attached, despite the fact that, as we have seen, it was hardly the first onboard mechanism to automatically graphically register information about a plane's performance.

While not the first wire recorder to be used by a flight crew, Warren's device was the first to be conceived and constructed as a permanent piece of aircraft apparatus, as well as the first to be expressly intended as an aid to accident investigation. Similarly, while not the first flight-data recorder (even the Wright Flyer had been equipped with instruments, crude though they were, to record such basics as airspeed, engine revolutions, and flight duration), Ryan's device was the first to truly emphasize in its engineering and design the concept of crashworthiness, as well as the first to be put through a rigorous program of scientifically controlled destructive testing. Hence, Warren's and Ryan's respective machines were more than mere flight recorders. They were forensic media.

Warren wrote *A Device for Assisting Investigation into Aircraft Accidents* in April 1954 in Melbourne, Australia. One year to the month later, at a meeting of the American Society of Mechanical Engineers in Baltimore, Maryland,

Ryan presented a paper on his new invention, the General Mills Ryan Flight Recorder. One scientific paper outlined an idea for a soon-to-be-built-and-tested cockpit voice recorder; the other delineated the technical features and practical functions of a recently-built-and-tested flight-data recorder. Though there is no indication in the historical record that Warren and Ryan knew of each other or of each other's research, their respective ideas and inventions were mutually entangled, not only nominally—each carrying the name "flight recorder" and, later, "black box"—but institutionally and ideologically. For we can see, in each of these technical artifacts from the mid-1950s, the crystallization of the cultural desire *to know the accident forensically* (find its cause, grasp its order, codify its meaning, divine its truth); to render it manipulable, narratable, docile, whether through mechanical reproduction (voices, sounds) or through mechanical representation (lines, curves). We can see here, as well, the crystallization of the cultural dream of making the accident not only instructive but productive—productive of a safer, less risky future. Both of these scientized media technologies promised to hasten progress by securing the nonrecurrence of high-speed crash and catastrophe.

Inscriptions Do Not Lie

In the introduction to *The Black Box: All-New Cockpit Voice Recorder Accounts of In-flight Accidents*, Malcolm MacPherson advises his readers that the twenty-eight cockpit-voice-recorder transcriptions he has compiled "are as dramatic reading as you are likely to find, because they are the minute-to-minute, unvarnished accounts of what actually happened."[95] The choice of words is telling: the transcripts are not simply dramatic *and* documentary; they are dramatic *because* documentary. MacPherson explains that his preoccupation with the final utterances of ill-fated pilots owes precisely to the transcripts' singular blend of spine-tingling action and ontological authenticity: "I first started reading the transcripts of CVR [cockpit-voice-recorder] tapes when I was living in Nairobi, Kenya, working as a correspondent for *Newsweek*. I never thought of myself as a ghoul for this unusual fascination. Far from it. I was able to picture the drama of what these transcripts contained, and they were unlike any I had ever read, seen on the screen, or watched in a theater. They were *real*, relived on the printed page."[96]

The transcripts' allure, on this view, their ability to stir the emotions and engage the imagination (in a manner that surpasses the mimetic offerings of stage and screen, no less), hinges on the "unvarnished" accuracy of the written record, which, in turn, hinges on the indexical authority of the cockpit voice recordings. That MacPherson is moved to underscore the reanimate realness of the transcripts ("*real*" and "relived"), but not that of the recordings, betrays a certain epistemological prejudice: it literally goes without saying that the tapes tell the truth, that the audio is, in a sense (a very *real* sense), immediate, *alive*. Only one hermeneutic operation is recognized here, one act of translation—namely, the one entailed in the move from sound recording to written word. There is no need, apparently, to demonstrate either the veracity of the recordings or the reliability of the recording medium. After all, the recordings themselves prove that the recorder was actually *there*, on the scene, running smoothly, in the very *presence* of the accident. For MacPherson, cockpit voice recordings are not so much *trans*criptions as they are *in*scriptions. And, evidently, black-box inscriptions do not lie.

MacPherson is certainly not alone in trusting the tale of the tapes. Indeed, it is now commonly supposed that the black box "holds the secrets of the crash."[97] The opening voiceover in "Witness to Terror," an episode of the Discovery Channel series *The New Detectives: Case Studies in Forensic Science*, illustrates the point:

> On a stormy Halloween night, a passenger plane begins a routine landing but what happens is far from routine. Investigators must determine what terrible chain of events brought it down nose first, killing everyone aboard. In the fog, a Canadian commuter plane aborts a landing maneuver. Seconds later, it plows full speed into the woods nearby. Was the accident caused by mechanical failure or a fatal lapse of judgment? The answers are preserved in the indestructible black box. Dutifully recording a plane's final maneuvers and the crew's last words, black boxes help investigators make sense of the fields of destruction.
>
> After a crash, a black box is often the only surviving witness to terror.[98]

This narration sensationalizes its subject matter, to be sure. But the notion that the black box is capable of bringing order to chaos, of answering the awful question of catastrophe, is by no means restricted to the realm of commercial television. In press reports and public commentaries, in government inquiries and technical analyses, in bureaucratic regulations and judicial decisions, the black box is routinely regarded and portrayed as an eminently reliable witness, "the Sherlock Holmes of [the] accident," the Rosetta Stone of "what went wrong."[99]

So convinced are accident investigators of the evidentiary value of flight recorders that they—and the nations, institutions, and enterprises that sponsor them—frequently go to extraordinary lengths to locate and retrieve them. In

February 1996, for example, a Boeing 757, leased by the Dominican Republic airline Alas Nacionales from the Istanbul-based Birgen Air, crashed into the Caribbean. Nicholas Faith, author of *Black Box: The Air-Crash Detectives—Why Air Safety Is No Accident*, writes:

> Financing the rescue and assembling the equipment required to salvage the boxes was a truly international effort. The money to contract the US Navy was raised by the United States, the Dominican Republic and the German and Turkish authorities, as well as Rolls-Royce, who had manufactured the engines.
>
> This story has everything—except, of course, a happy ending. The aircraft climbed to around 7,000ft and ended up 7,200ft below the surface of the ocean, so finding and recovering the black boxes required the use of some very specialized equipment by the US Navy. Luckily, they were able to bring up the recorders on their first dive. The aircraft had disintegrated at the precise point where the recorders were located, which freed both of them and made them easier to extract than would have been the case had they remained attached to a much larger piece of the wreckage. It also helped that the DFDRs [digital flight-data recorders] were mounted at the back of the plane and had underwater locater beacons attached to them which emitted audible signals that could be detected from equipment in the salvage ships.
>
> Amazingly, the whole operation took only a single day, albeit an exceedingly long one.[100]

Is it any wonder that the hunt for the black box has been likened to "the quest for the Holy Grail"?[101] Faith's account focuses attention on the huge and complex infrastructure—social, industrial, political, technological—required to mount a major air-crash search-and-rescue operation. And for what, this elaborate undertaking? Not for the 189 souls aboard Birgen Air Flight ALW301; they were presumed dead. No, the technical knowledges and capital sums, the military personnel and "very specialized equipment," the government officials and deep-sea divers were all marshaled in an effort to salvage a couple of sunken metal boxes.[102] And why? Because, as *The New Detectives* narrator puts it, "the only way to determine what really happened [is] to look inside the plane's black box."[103]

If the amount of press coverage devoted to accident investigations is any indication, the lay citizen, as much as the aviation professional, gives credence to the technological testimony of the black box.[104] While airline customers, in particular, might be expected to take a keen interest in what flight recorders have to say (so to speak), their detached curiosity pales in comparison to the profound affective investment of crash victims' families.

In 1996, the *New York Times* interviewed relatives of those who perished in that year's Valujet and Trans World Airlines accidents:

> Susan Smith says she knows there are people who might not understand why she wants to hear the tape of the final terrifying minutes of the Valujet flight that caught fire and crashed in the Everglades in May, killing her 24-year-old son, Jay, and the other 109 on board. According to a transcript of the tape, which has been released, the passengers were screaming that the plane was on fire.
>
> "If it hadn't happened to me, I might have felt, 'Well, I wonder why they want to hear it,'" Mrs. Smith said. . . . "I'd think that I wouldn't want to hear the horrific things that are on the tape. But when it's someone you love, you want to hear every detail to put every puzzle piece together. You want to know everything. Nothing seems morbid anymore.
>
> "My son wasn't protected," Mrs. Smith added. "Why should I be protected?" . . .
>
> Marilyn Chamberlin, the mother of the Valujet pilot, 35-year-old Candalyn Kubeck, said in an interview that she wanted to hear her daughter's voice one last time. "A mother can tell better than anyone her state of mind, how frightened she was," Mrs. Chamberlin said. . . .
>
> Laura Sawyer is one of those who wants to hear the tape of the Valujet flight. Her grandparents . . . were on the plane. . . . "My grandfather was extremely loud," said Ms. Sawyer. . . . "He was very, very hard of hearing. You could hear him across a room. As morbid as it might sound, I am really interested in hearing what their last words were."[105]

On the one hand, these remarks express the fraught, complicated dynamics between the grieving process and the will to knowledge, where the need to know implies not only the impulse to obtain the "objective" facts but also the compulsion to identify with, or even vicariously experience, the dread and distress of a loved one. As Constance Penley argues, "The desire to know, no matter how shot through with fantasmatic thinking—disavowal, guilt, projection, overidentification, and all the rest—is a desire whose reality and efficacy must be acknowledged. The families of MIAs, of accident and murder victims, even the relatives of those killed by Jeffrey Dahmer want to know, demand to know, in exact and excruciating detail, how their loved ones died. Over and over, they say that they cannot go on, cannot resume their own lives until *they know*. Tell us anyway, they say, no matter how horrible; knowing is infinitely better than not knowing."¹⁰⁶ On the other hand, the crash victims' family members' poignant comments suggest the black box's popular epistemological authority, its status as a trusted reproducer of indexical reality. Preserve and play back the substance of the real—does anybody doubt that this is what recording media really do?

Not Present, but in the Voice! (A Brief Interlude)

Thomas Edison's principal marketing agent in Europe was Colonel George Edward Gouraud. Having "played a major role in the European introduction of [Edison's] telephone and lighting system," Gouraud in the late 1880s, according to John Picker, "became the most vocal phonograph enthusiast in Britain, recording dozens of prominent and obscure Victorians for posterity."¹⁰⁷ To promote the first demonstrations in London of Edison's new and improved phonograph of 1888, Gouraud proposed to issue invitation cards to distinguished personages that read:

> To meet Prof. Edison,
> Non presentem, sed alloquentem!¹⁰⁸

Edison would not be present *in the flesh* at the London demonstrations, but he would be there on wax cylinder, *alloquentem*, "in the voice!"

Parasite

In classical information theory, noise is the enemy to be eliminated. Optimal communication between two poles—the encoder or sender on one side, the decoder or receiver on the other—is achieved when signals are maximized and random disturbances (interference, static, noise) are minimized. The greater the polar isomorphism, the more felicitous the informational exchange. Communication is a matter of purifying the channel, of mathematically separating message from clutter.

Michel Serres spins the story differently—and, in a way, so does the black box. There is no signal without noise, Serres insists, no pure and clean transmission, no free and clear exchange of information, much less of "meaning." It is not enough, however, to recognize, as does Claude Shannon, that communication is an inherently stochastic process. Neither is it sufficient to draw attention, as does Norbert Wiener, to the entropic forces that threaten the integrity of any homeostatic system. Such claims, for Serres, must be reimagined, radicalized, their valuation, in effect, inverted. Shannon's random disturbances are not impediments to exchange; Wiener's entropic forces are not threats to the system. Or, rather, they *are* impediments and threats, but they are also—because of this, not in spite of this—the very condition of possibility of exchange and systematicity. "Noise. . . . One parasite (static), in the sense that information theory uses the word, chases another, in the anthropological sense. Communication theory is in charge of the system; it can break it down or let it function, depending on the signal. A parasite, physical, acoustic, informational, belonging to order and disorder, a new voice, an important one, in the contrapuntal matrix."[109] Noise, or what Serres, playing on the double meaning of the French word, calls "parasite," is not "something added to a system, like a cancer of interceptions, flights, losses, holes, trapdoors"; instead, "it is quite simply the system itself."[110]

As a test of his then-unbuilt device's in-flight potential, David Warren, in 1957, took a Minifon recorder up to the skies. He soon discovered a daunting problem: "The background noise of the aircraft all but drowned out the pilot's conversation."[111] Sound-recording media "do not only store . . . the one signified, or trademark, of the soul," Friedrich Kittler notes. "They are good for any kind of noise."[112] In a bid to exterminate the soulless trouble, Warren "experimented by designing filters that would delete the surrounding noise, yet would clearly record the voice."[113] But the engine's din proved a recurrent interference, a persistent pest. Ultimately, the best Warren could do was to optimize "the speech-to-noise ratio."[114]

Nowadays, when an airliner crashes, an official committee, typically composed of members from the National Transportation Safety Board, the Federal Aviation Administration, the aircraft operator, the airplane manufacturer, the engine's manufacturer, and the pilots' union, is convened to listen to the cockpit voice recording.[115] The members of this "CVR committee" listen attentively, diligently, to the playback of the flight crew's conversations, over and over, often for weeks on end, taking note of exactly what the pilots said and exactly when they said it. And so commences a series of translational operations, most of whose relays or shifts in register, like the one entailed in the move from sound recording to written word, are elided or obscured. Auditory impressions are received . . . the words of the flight crew are heard (that is, the auditory impressions are registered *as words*) . . . the words of the flight crew are interpreted . . . the interpretations are used to ascertain the accident's "true" cause . . . the cause of the accident is announced to the public. . . .

Often these interpretations are popularly reproduced as dramatizations, as heroizations—heroizations of the pilots, of course, but in some cases, such as

United Airlines Flight 93, which crashed in Pennsylvania on September 11, 2001, of the passengers as well.[116] MacPherson contends that cockpit voice recordings "can go a long way to explain in the most graphic and dramatic terms the professionalism, if not the outright heroism, of many commercial pilots and crews and the extraordinary measures they take to ensure our safety."[117] The forensic-audio analyst Mike McDermott concurs: "In the vast majority of the tapes I've listened to where professional pilots are encountering difficulties which result in a crash, they are working very hard to prevent that crash all the way down to the ground. It could be that they're going straight down and at 35ft from the ground they're still looking for something to do to save the aircraft. Generally, they do not panic."[118]

Even as they listen to the professional pilots' speech, cockpit-voice-recorder experts like McDermott listen for other, "accidental" sounds, random disturbances "in the contrapuntal matrix," blips and buzzes, "losses, holes, trapdoors," clatter, clutter, noise. As the journal *American Heritage of Invention and Technology* points out, "More than a few accidents have been solved by CVRs, but not, as one might suppose, from hearing what the pilot and copilot are saying, which is often no more revealing than 'What's that?' or 'That looks odd.' Rather, the detective-like engineers at the NTSB decipher and analyze the variety of clicks, rattles, and engine hums behind the voices."[119] Similarly, the *New York Times* reports that "the machine noise often proves invaluable in analyzing accidents."[120]

Just as Serres said: noise, static, is always already part of the signal. The noise is *necessary*. Contra Aristotle, accidents cannot be expelled, cannot be banished from the system: they keep coming back—and never really went away. "We know of no system that functions perfectly, that is to say, without losses, flights, wear and tear, errors, accidents."[121] Noise haunts the signal as death haunts life. But also: noise haunts life as death haunts the signal. Jonathan Sterne has coined the term *audile technique* to designate a modern "set of practices of listening that [are] articulated to science, reason, and instrumentality and that encourage[] the coding and rationalization of what [is] heard."[122] Trained to scientifically rationalize and codify their own auditory impressions, cockpit-voice-recorder experts deploy a forensic version of audile technique as they listen to reproductions of human speech and ambient sound. What they hear, at the same time, and nevertheless, is death. Or, rather, they hear life *disturbed* by death—*that* is the noise.

For Kittler, the age-old dream of recording media, in the aftermath of Edison's talking machine's invention, "becomes at once reality and nightmare."[123]

Technical media are necessarily "engulfed by the noise of the real—the fuzziness of cinematic pictures, the hissing of tape recordings."[124] On their material surfaces, the real manifests (symptomatically) as unilluminable excess: blurry traces, sibilant transmissions, unruly corporalities. "Of the real nothing more can be brought to light than what Lacan presupposed—that is, nothing. It forms the waste or residue that neither the mirror of the imaginary nor the grid of the symbolic can catch: the physiological accidents and stochastic disorder of bodies."[125] Audio distortions, video dropouts, online lags, compression artifacts: such technological nuisances, asserts Brian Larkin in *Signal and Noise*, "are the material conditions of existence for media."[126]

The black box is tainted with darkness and debris, chaos and cacophony, misfortune and finitude. It reeks of death, rattles with death (the noise again). Death informs its technical function, its cultural meaning, its institutional purpose. Yet, at the same time, its written transcripts, MacPherson assures us—even they, those copies of copies—are so really *alive*! The recordings themselves, all the more so! Thus, Susan Smith "wants to hear the horrific things that are on the tape," no matter how morbid her wish might seem. Marilyn Chamberlin desires to know the degree of her daughter's terror by detecting the cadence and timbre of her mechanically mummified voice. Laura Sawyer hopes to somehow hear the last words of her hard-of-hearing grandfather. And when they listen—listen to the device's listening-in—what will they hear? *Death sucking on the signal*: the parasite. They will hear the flight crew's desperate attempts to regain control of the out-of-control airplane, to intervene in and counteract the worsening crisis, to "chase the parasite." Alas, to no avail. For, despite the pilots' last-ditch efforts to leverage all of their professional skill and experience, to bring illumination to the darkening drama, "a new obscurity accumulates in unexpected locations, spots that had tended toward clarity; [they] want to dislodge it but can only do so at ever-increasing prices and at the price of a new obscurity, blacker yet, with a deeper, darker shadow. Chase the parasite—he comes galloping back, accompanied, just like the demons of an exorcism, with a thousand like him, but more ferocious, hungrier, all bellowing, roaring, clamoring."[127]

The parasite, noise, death: it can never be chased *out* because without it there is no *in* in the first place. It keeps returning, and it never really departed. In the cockpit, it is there and not there (a fly on the wall, a bug). On the cockpit voice recordings, it hums residual presence and hisses excessive absence, "reality and nightmare." *Non presentem, sed alloquentem!* The relatives of plane-crash victims want so badly a playback demonstration. But what, finally, will they hear on this "family record"? Nothing but a leech feeding on a fleshless host.

How, then, does the black box indicate that death has arrived? What sound, what noise could possibly betoken this overcoming coming? This ultimate presenting that is also an ultimate absenting? This final, terrible once-and-for-all? How will Susan Smith and Marilyn Chamberlin and Laura Sawyer, who wish to know, who *need* to know "in exact and excruciating detail," as Penley says—how will they know for sure when death has come to claim their loved ones? For, in the case of the black box, death does not "come knocking." Neither does it come clicking, as it does in the case of the telephone. Avital Ronell, in her chapter titled "The Black Box: After the Crash: The Click: The Survival Guide," writes, "The Bell telephone shapes a locus which suspends absolute departure. The promise of death resisted, however, destines itself toward the click at your end. The click, neither fully belonging to the telephonic connection nor yet beyond or outside it, terminates speech in noise's finality. A shot that rings out to announce, like an upwardly aimed pistol, the arrival of silence . . . the click stuns you."[128] The end of all speech, of all sound, of all signal, of all noise too, comes to the black box differently. The instant of irrevocable disconnection, of unbearable overdetermination, arrives not with a click that rings out like a bullet, but with the sudden, startling onset of silence—that is, with nothing. Warren, the first to give it a name, called it "immediate cut-out." The sonic sign of death is here no sign at all; the sign of death is the terminal silencing of signification itself.

How to visually and verbally represent this absolutely unrepresentable ending? This endless caesura? In his collection of CVR transcripts, MacPherson renders the black box's decisive yet undecidable moment by skipping a couple of lines after the last transcribed words (or vocalizations or automated commands). Centered on the page in boldfaced capital letters appears the phrase "end of tape." Like this:

END OF TAPE

This simple typographic technique is surprisingly (and uncannily) effective, although for maximal impact, as it were, the phrase "end of tape" must be read as operating "under erasure." Hence, it would be better to represent the black box's immediate cut-out like this:

~~END OF TAPE~~

For Jacques Derrida, a term or phrase "under erasure" is one that is simultaneously inadequate to and necessary for a given linguistic or conceptual purpose.[129] The typographic sign **~~END OF TAPE~~** captures several key aspects of its signified (non)phenomenon at once. The sign's presence on the page marks the trace of a terminus. Its placement on the page (two lines below, centered) marks the trace of a separation and of a concentration. Its capitalization marks the trace of an emphasis. So does its boldface, but this augmented blackness also marks the trace of an inscrutable darkness, a non-illumination. Its concise phrase marks the trace of a truncation. Its functionality marks the trace of a generic convention, a literary closure: the old-fashioned story or fairytale that concludes, typographically, with "THE END." These are the sign's necessities. Its strikethrough, however, points to the sign's irreducible inadequacy, its impossible signification: it marks the trace of a parasite.

Last Words

In his *North American Review* article, Edison claimed that, "for the purpose of preserving the sayings, the voices, and *the last words* of the dying member of the family—as of great men—the phonograph will unquestionably outrank the photograph."[130] One year earlier, in 1877, Edward H. Johnson, writing under his own name in *Scientific American*, declared,

> It has been said that Science is never sensational; that it is intellectual not emotional; but certainly nothing that can be conceived would be more likely to create the profoundest of sensations, to arouse the liveliest of human emotions, than once more to hear the familiar voices of the dead. Yet Science now announces that this is possible, and can be done. . . . Whoever has spoken or whoever may speak into the mouthpiece of the phonograph, and whose words are recorded by it, has the assurance that his speech may be reproduced audibly in his own tones long after he himself has turned to dust. The possibility is simply startling. A strip of indented paper travels through a little machine, the sounds of the latter are magnified, and our great grandchildren or posterity centuries hence hear us as plainly as if we were present. Speech has become, as it were, immortal.[131]

The cultural history of the cockpit voice recorder begins here: with the Edisonian family record; with the startling immortality of speech; with the virtual presence of voices of the dead; with what John Durham Peters calls the phonograph's "mausoleum of sound, fixed in a state of suspended animation"; with the mechanically preserved words of souls now departed, of kindred gone to dust.[132]

Or perhaps that history begins much earlier. In promoting the phonograph as a practically magical repository for the storage and safekeeping of "the last words of the dying," Edison was cleverly tapping into a long-standing and abiding transcultural fascination with final spoken expressions, with things uttered in extremis. As near-death moments are often charged with awe and expectation, so the oral communications (or attempts threat) of the soon-to-be-dead are widely supposed to be laden with significance, even brimming with mystical or quasi-mystical meaning. "Particular importance is, and has always been, attached to last words," writes Karl Guthke in his study of the subject. "Indeed, they have been treasured since time immemorial in cultural communities that otherwise have little in common."[133]

Traditionally, end-of-life speech has been accorded a privileged status, a "special testimonial value," in realms both spiritual and secular, both institutional and personal.[134] Insofar as the ancients believed in the oracularity of last words, it was because they likewise believed, as Socrates avers in Plato's *Apology*, that "the gift of prophecy comes most readily to men—at the point of death."[135] Medieval Christianity dramatized the deathbed as a place of eleventh-hour repentance, the scene of the sinner's last chance to plead for divine forgiveness, to gain salvation through verbal confession or affirmation of faith.[136] Western literature and drama, as Guthke's research amply demonstrates, have "made the widest possible use of last words—this high-brow folklore that surrounds us everywhere."[137] In Anglo-American jurisprudence, meanwhile, "one of the oldest exceptions to the hearsay rule"—to wit, the rule of evidence barring the admission of out-of-court statements made by persons unavailable for cross-examination—"is embalmed in the phrase 'dying declaration,'" according to the *Howard Law Journal* contributor Charles Quick. "As a matter of fact, the admission of dying declarations preceded the development of the hearsay rule," stretching back to the English courts of the early thirteenth century, wherein, as Brendan Koerner observes, "the principle of *Nemo moriturus praesumitur mentiri*—a dying person is not presumed to lie—originated."[138] Still in effect to this day, the hearsay exception for dying declarations ratifies the premodern notion, rooted in supernatural belief and religious custom, that last words, issuing as they do from the mouth of someone about to meet his or her Maker, of a soul dangling precariously between life and afterlife, between the surly bonds of earth and either the pearly gates of heaven or the punishing fires of hell, tend to be especially sincere words, genuine and trustworthy words, *true* words. Finally, as Robert Kastenbaum notes—and as Thomas Edison evidently recognized—"the last words of people in everyday life" can reverberate for years to come in the minds and memories of the decedent's relatives and descendants. "Family traditions may include reminders that 'This is what Father asked of us from his deathbed,' or 'Mother's very last words were. . . .' Statements of this type function as instructions from the deceased whose moral and emotional power is enhanced by the circumstances in which they were imparted."[139]

Cockpit voice recordings do not, as a rule, circulate publicly. Various snippets, many of them quite harrowing, can be streamed from websites such as AirDisaster.com, AviationExplorer.com, and PlaneCrashInfo.com, as well as YouTube.com.[140] However, in the United States, the National Transportation Safety Board is prohibited by law from publicly disclosing or disseminating any audio recording "of oral communications by and between flight crew members and ground stations related to an accident or incident investigated by the Board."[141] Besides the airline companies, which own the original recordings, usually only the courts and the families of crash victims are granted authorized access—and then only conditionally—to such sensitive material, such anguishing sounds and utterances. In general, only those who *officially* need to know, on the one hand, and those who *emotionally* need to know, on the other, have the duty or the opportunity to be "in the know." Their ears alone are allowed to apprehend the last words preserved (as though for posterity) on the CVR, to receive "the sayings, the voices" entombed (as though for eternity) in the black box.

Seven months after the September 11 terrorist attacks, some seventy relatives of United Flight 93 passengers and crew gathered in a hotel ballroom in Princeton, New Jersey, to listen to thirty-one minutes of CVR audio—audio that was thought to hold "special testimonial value." *The possibility is simply startling.* "They came hoping to glean a voice, a sound, a telltale hint of how the last moments of their loved ones were spent," reported the *Pittsburgh Post-Gazette* on 19 April 2002.[142] And when they listened, what did they hear? Alice Hoagland, mother of posthumously celebrated passenger Mark Bingham, "said the voices were 'very poor quality,' often muddled and occasionally drowned out by the sound of wind rushing outside the cockpit as the plane sped at 575 mph at a low altitude. The background noise made it difficult to discern individuals 'even when people were yelling at the top of their voices.'"[143] Once more, the noise. Before Hoagland and the other auditors donned their earphones that day, the Federal Bureau of Investigation, which had organized the strange and somber gathering, advised them that what they were about to hear was "violent and very distressing." *A little machine creates the profoundest of sensations, arouses the liveliest of emotions.* When the black box is experienced in its full intensity, in its rude immediacy, the Bureau warned, "it may be impossible to forget the sounds and images it evokes."[144]

There are fewer legal restrictions on CVR transcripts than on the recordings themselves. In fact, U.S. law stipulates that the NTSB "shall make public any part of a transcript or any written depiction of visual information the Board decides is relevant to the accident or incident."[145] Because the NTSB is empowered to exercise discretion as to which parts of a given CVR transcript are "relevant to the accident," and because other national air-safety bureaucracies are normally similarly empowered, the government-authorized transcriptions compiled in books such as MacPherson's *The Black Box: All-New Cockpit Voice Recorder Accounts of In-flight Accidents* and Marion Sturkey's *Mayday: Accident Reports and Voice Transcripts from Airline Crash Investigations* are not only partial but, indeed, preemptively pruned documents. While not heavily censored, they are nevertheless officially sanitized versions of "what really happened." On top of this, both MacPherson and Sturkey further edited and abbreviated their respective selected transcripts prior to publication. "This book does not include the complete transcripts," Sturkey explains in the prologue to *Mayday*. "The portion of transcripts quoted in this book generally begins just before the pilots encounter the dilemma that leads to the accident. Thereafter, non-pertinent cockpit noises and non-pertinent conversation are excluded in the interest of brevity."[146] Yet the Bowdlers at the NTSB are occasionally imperfect or inconsistent expurgators, and even in Sturkey's compilation (which, unlike MacPherson's, refrains from printing expletives) some "non-pertinent conversation"—some linguistic and some paralinguistic "noise"—apparently slipped through the cracks.

Sometimes the last words stored on a doomed airliner's cockpit voice recorder constitute a simple, if urgent, imperative, an order or instruction from one pilot to another about what to do next, what emergency measures to take. Sometimes the last words (or whoops) are automated alerts or commands, "advisory callouts" from the plane's ground-proximity-warning system made the more eerie by their intonational uniformity and robotic impassivity: *Too low—terrain. Whoop. Whoop. Pull up. Whoop. Whoop. Whoop. Terrain. Terrain. Pull up. Pull up.* Sometimes the last words are interjections, a crew member's curse or exclamation, swearword or fear scream: "Damn it!" "No!" "Oh, God!" "Aaahhh!" "No way!" "Whoa!" "Shit!"[147] (Recall the Australian Air Force's prediction that David Warren's device probably "would yield more expletives than explanations.") Sometimes the last words are expressions of bafflement, pure confusion: "Oh, what's happening?" "What the hell is this?" "What's going on now?" "I have no idea which way is up!" Sometimes the last words are declarations of imminent self-demise, a statement of fatal resignation often phrased as a fait accompli—death proleptically announcing its ineluctable presence: "That's it, I'm dead." "Ah, we're finished." "We're over." "We're dead." Sometimes the last words are sung rather than spoken or shouted: according to MacPherson, one of the pilots of Turkish Airlines Flight 981, which crashed near Paris in 1974, warbled a child's lullaby as the plane went down.[148] And sometimes the last words are hauntingly, heartbreakingly intimate goodbyes, a final farewell to a relative or romantic partner from an aviator who suddenly realizes he or she is not long for this world: "Ma, I love yah!" "You got it all, Dad! We're gonna hit!" "Amy, I love you." In such rare and remarkable instances, the pilot, having abandoned all hope of survival, not only directly addresses the cockpit voice recorder (the lurking existence of which had, until that moment, been either forgotten or willfully disregarded in accordance with aviation norms), but knowingly appropriates it as a medium of deeply personal, *desperately* personal, communication.

The tragic voices of those about to die, the pathetic vocalizations of the already dead: these, as Kittler says, are "what media can store and communicate." After the crash, before and beyond immediate cut-out, the black box contains all—and nothing—of "what remains of people."

four

TESTS AND SPLIT SECONDS

One could argue that, nowadays, . . . there is nothing that is not tested or subject to testing. We exist under its sway, so much so that one could assert that technology has now transformed [the] world into so many test sites. Among other things, this means that everything we do is governed by a logic that includes as its necessary limit, probability calculation, self-destruction, and interminable trial. —AVITAL RONELL, "The Test Drive"

Film makes test performances capable of being exhibited, by turning that ability itself into a test. —WALTER BENJAMIN, "The Work of Art in the Age of Its Technological Reproducibility: Second Version"

The complicated physical and physiological events involved in automobile collisions were little examined and little understood prior to the Second World War. The car crash's "natural" phenomena—its energies, its materialities, its spatialities and temporalities—were not generally considered worthy of serious scientific inquiry. Ideologically and institutionally, the focus in the early decades of automobility in the United States was on preventing (and thereby eliminating) accidents, not on discovering what took place during their occurrence. What mattered to safety experts and safety advocates of the day was the moral and behavioral reformation of the so-called reckless driver, not the systematic investigation of "vehicular collision dynamics."[1]

By the early 1950s, however, the basic terms and tendency of auto-safety discourse had changed dramatically. The emphasis was no longer on the prevention (and progressive elimination) of accidents, but rather on the reduction of crash injuries and fatalities. This conceptual and discursive shift—from a regime of crash avoidance to one of crash amelioration—was tied to the emer-

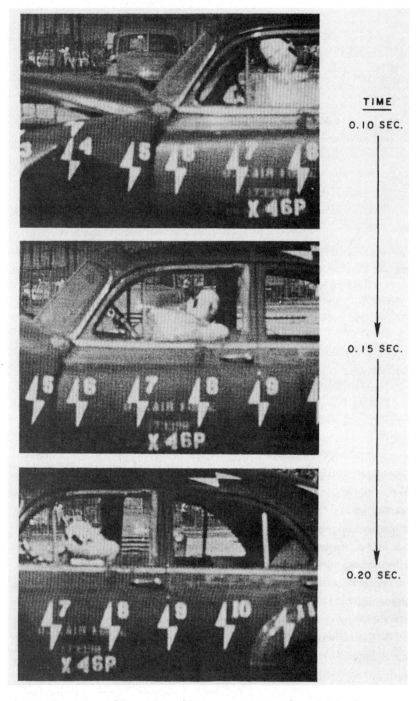

Photographic analysis of the crash's awful moment. Courtesy of UCLA's School of Engineering and the Regents of the University of California.

gence of a new technoscientific ritual: the automobile collision experiment. Full-scale "accidents," complete with humanoid dummies as human surrogates, were painstakingly re-created at several industrial and institutional sites across America during the postwar period. Intricate systems of instrumentation, electronic and photographic, were used to facilitate the observation, registration, and analysis of the collision process in all of its aspects: every motion, every mechanical deformation, every anatomical contortion. High-speed cinematography, in particular, held the promise of new observational and analytical knowledge. With its ability to visually apprehend and temporally extend instantaneous activities—to seize them, slice them up, and slow them down—the overcranked camera enabled unprecedented perceptual access to the crash's most eventful, and awful, moment.

Postwar auto-safety discourse invested that eventful, awful moment with profound urgency and significance. It stressed that everything truly consequential about automobile catastrophe was to be found therein: the excessive forces, the uncontrolled actions, the structural failures, the deadly blows. If crash injuries and fatalities were to be reduced, so the logic went, the instant of impact would have to be mined for information, studied in depth and detail, revealed and rationalized. Long considered terra incognita, the destructive split second was now, for the first time in automotive history, territory that demanded to be explored and explained, and high-speed cinematography offered collision experimenters of the 1950s a unique and compelling means of mapping and making sense of it. Behind this peculiar scientific cinematographic project lay a broader cultural desire to turn the accident of accelerated mobility, that recurrent source of dread and disruption in modern life, into an object of forensic knowledge and institutional control.

Blaming the Beast

There is a moment early in the history of every transportation technology when the terror of the accident has not yet reached critical mass in the collective consciousness. In such moments, the new mode of transport is hailed as more efficient, more reliable, and more secure than the older, conventional mode; it arrives as an improvement, a "civilized" refinement. Over the course of the nineteenth and early twentieth centuries, the horse-drawn carriage was largely replaced and rendered obsolete, first by steam engine and railway, then by internal-combustion engine and roadway. In each case, the newer mechanized vehicle was thought superior, at least initially, to the more traditional animal-powered vehicle in terms of ease, safety, and overall dependability.

"The promoters of the railroad regarded steam power's ability to do away with animal unreliability and unpredictability as its main asset," observes Wolfgang Schivelbusch. "Mechanical uniformity became the 'natural' state of affairs, compared to which the 'nature' of draught animals appeared as dangerous and chaotic."[2] Schivelbusch quotes an anonymous text published in England in 1825:

> It is reasonable to conclude, that the nervous man will ere long, take his place in a carriage, drawn or impelled by a Locomotive Engine, with more unconcern and with far better assurance of safety, than he now disposes of himself in one drawn by four horses of unequal powers and speed, endued with passions, that acknowledge no control but superior force, and each separately momentarily liable to all the calamities that flesh is heir to. Surely an inanimate power, that can be started, stopped, and guided at pleasure by the finger or foot of man, must promise greater personal security to the traveller than a power derivable from animal life, whose infirmities and passions require the constant exercise of other passions, united with muscular exertion to remedy and control them.[3]

Similar claims were made on behalf of the automobile at the end of the nineteenth century. In 1896, the New York-based magazine *Horseless Age* promised that "the motor vehicle will not shy or run away. . . . Frightful accidents can be prevented. The motor vehicle will do it."[4] The next year, the pioneering American automaker J. Frank Duryea put it this way: "The horse is a willful, unreliable brute. The ever recurring accidents due to horses which are daily set forth in the papers prove that the horse is a dangerous motor and not the docile pet of the poet. The mechanical motor is his superior in many respects, and when its superiority has become better known his inferiority will be more apparent."[5] Here is *Horseless Age* again, this time in 1899: "[The] truth is, we are just beginning to realize what a fractious, unreliable animal the horse is. . . . The animal is treacherous and dangerous, and his gradual elimination from the centers of civilization is not only much to be desired, but . . . a necessity."[6] Finally, that same year, *Harper's Weekly* had this to say on the subject: "A good many folks to whom every horse is a wild beast feel much safer on a machine than behind a quadruped, who has a mind of his own, and emotions which may not always be forestalled or controlled."[7]

Underlying such assertions is the notion that accidents result from the impossibility of ever fully taming or domesticating the animal, on whom the efficient movement of people depends. Harm and hazard arise because the beast's urges and wild impulses can be neither anticipated nor wholly suppressed, because its every move cannot be determined or precisely managed by its nominal master. Human bodily injury and psychic insecurity are the inevitable consequences of lower creatures' natural infirmities and intractabilities. Animals "bite and buck and act in ways that often cannot be controlled by humans," as Sarah Lochlann Jain notes. "They hurt people and as such [are often] described as vicious and bad."[8] Such brutes can be brought into provisional submission, perhaps, but their basic instincts can never be beaten out of them; their flesh cannot be made less treacherously fickle. Hence, when horse-drawn travel turns deadly or disastrous, blame falls on the beast.

"Man," on this conception, is above reproach. Mishaps are visited on him, not caused by him; disasters occur despite him, not because of him. His technological marvels, railroad train and automobile, are not the problem. Products of reason and enlightenment, instantiations of civilization and social progress, these reliable, predictable machines promise to rid accelerated mobility of its many perils. No more irregularities of motion or paroxysms of animal passion. No more desperate struggles to control the source of motive power. No more shying or running away, biting and bucking. No more willful misbehavior. No more danger or chaos. No more "frightful accidents."

Such claims are soon enough shown to be specious, the fancies of boosters, dreamers, and hucksters, the inflated hopes of those who want or need to believe. As accidents accumulate, one after the other, each seemingly more violent and destructive than the last, it becomes increasingly difficult to credibly sustain the aura of safety that originally surrounded the new transportation technology. As the death toll continues to climb, day after day, with no end in sight, the rhetoric of efficiency and dependability begins to ring hollow. The idea, once so luminous, that the machine is substantially more secure than the animal, loses its luster a little more with each new railroad "concussion" or automobile collision. And once the luster wears off completely, the idea is liable to wobble and overturn: the machine is feared and condemned for the scale, suddenness, and sheer violence of its mishaps, while the animal is remembered fondly for the peace of mind it brought the preindustrial traveler, who took comfort in the creature's gentle carriage and closeness to nature.

By the first decade of the twentieth century, studies by the insurance industry indicated that the motorcar was more menacing than the horse-drawn coach: "Automobile accident statistics released at the 1909 meeting of the International Association of Accident Underwriters revealed that the motor vehicle was not safer than the horse, for the record was much worse than what had been considered normal for horse-drawn vehicles. In the same year, *Horseless Age* admitted that 'the "automobile hazard" is not likely to decline in frequency.' Not only did the accident rate seem very high given the number of automobiles in use, but automobile accidents usually involved more serious bodily injury and higher property damage claims than accidents involving only horse-drawn vehicles."[9] Thus arose a new dilemma. With the beast no longer blamable, who or what was responsible when something went wrong, when mobility went amiss? Joel Eastman, author of *Styling vs. Safety: The American Automobile Industry and the Development of Automotive Safety, 1900–1966*, states that "because the automobile, unlike the horse, was an inanimate object, it was natural for the blame for accidents to be laid on the human operator."[10] And there it lay, by and large, for the first half of the twentieth century.

Accident Discourse in the Early Decades of Automobility

From the earliest days of automobility in the United States, the notion held sway that traffic accidents were almost always the product of driver negligence. A newspaper reporter in 1902 wrote, "For the automobile has not, like a horse, a will of its own, which may act uncertainly. It is sensitive and responsive—acting in exact accordance with the principles upon which it is constructed. The accident which happens to the automobile is seldom due to the machine itself, but almost wholly to the loss of control or presence of mind of the operator."[11] Notice the peculiar phrasing here: it is the automobile, not the operator, to which the accident happens. "Man," once the irreproachable victim of the animal's unreliability, has himself become the unreliable animal, even as the machine, once and still the embodiment of safety, has become the irreproachable victim. The beast is still to blame, but now the beast is human.

In general, the law courts of the day adhered to this "driver negligence paradigm," to borrow Jain's term.[12] A precedent-setting judicial opinion issued in 1907 declares: "It is not the ferocity of automobiles that is to be feared, but the ferocity of those who drive them. Until human agency intervenes, they are usually harmless."[13] François Ewald, writing in a Foucaultian vein, describes this "juridical logic of responsibility": "The judge takes as the point of departure the reality of the accident or the damage, so as to infer the existence of its cause in a fault of conduct. The judge supposes that there would have been no accident without a fault. . . . Juridical reason springs from a moral vision of the world: the judge supposes that if a certain individual had not behaved as he or she actually did, the accident would not have happened; that if people conducted themselves as they ought, the world would be in harmony."[14] If only the person behind the wheel would be more careful and conscientious, pay more attention, play by the rules. If only he or she would not behave so ignorantly, so irresponsibly. Under the juridical logic of responsibility, mishaps are avoidable and preventable, and their occurrence is attributable to the abnormal actions—the "reckless" conduct—of motorists. The problem of the accident is defined as a culpable lack of driver discipline. The driver who gets into a wreck is undisciplined socially (unadjusted) or perceptually (unfocused) or mentally and manually (untutored, untrained). Above all, as Ewald suggests, he or she is undisciplined morally (unredeemed). His or her recklessness is a sign of selfishness, an expression of blatant disregard for the rights and well-being of others. He or she is a wrongdoer, a malefactor, a "bad" driver.[15]

The automobile industry strongly endorsed this line of reasoning. It had an obvious interest in doing so, as such reasoning served a threefold purpose: to absolve it of moral responsibility; to exempt it from legal liability; and, not least, to ward off the specter of government regulation.[16] Pointing the finger at the reckless driver made the industry appear inculpable, as did pointing the finger at poor road conditions, the industry's other favored explanation for why accidents happened. Not incidentally, these strategies of blaming implied that there was no need to reengineer the automobile for safety. Why should automakers waste (or be compelled to waste) their valuable time, energy, and resources on elaborate and expensive safety research and development when accidents were not their fault? Besides, what good would redesigning the vehicle do if all the while the driver remained unredeemed? "The industry argued that automobiles almost never cause accidents and that it was not normal for one to be involved in a crash—thus, there was no obligation to design a car with this possibility in mind. It was vociferously maintained that even the 'perfect vehicle' could not prevent driver error or poor road conditions, and that, thus, the solution to the safety problem lay in improving the driver and the highway."[17]

Fortunately for the automobile industry, the highway-safety movement, a loose aggregation of public and private organizations that emerged around 1914, the most important of which was the National Safety Council, institutionalized and helped to popularize the idea that traffic accidents were caused by bad drivers and bad roads, not bad cars. "The rationale for the Council's approach to accidents was based on the assumption that all accidents had causes; as one author put it: 'Human misbehavior, human frailty, human ignorance, human laziness all cause accidents. Remove them and accidents cease.'"[18]

The guiding philosophy of the highway-safety movement was summed up in the slogan known as "The Three E's: Engineering, Enforcement, and Education."[19] The problem of the accident was to be solved through a strategy of triangulation, with each term indicating a distinct approach and angle of attack. The first called for improvements in roadways and related infrastructure;

the second, for the enactment of new traffic laws and the establishment of new traffic courts; the third, for the mass dissemination and public inculcation of "safety-first principles." According to Eastman, "The safety professionals concluded that, since almost all accidents could be attributed to some human action—which was usually in violation of at least one traffic ordinance—the solution was to educate drivers and pedestrians to behave 'safely' and legally and to enforce the laws against those who misbehaved."[20] Significantly, those same safety professionals had little to say about the automobile, other than to praise it as "a nearly perfect mechanism which caused few, if any, accidents."[21]

The combined efforts of the automobile industry and the highway-safety movement to instill a particular understanding of—and to naturalize a particular discourse about—the relationship between driver (reckless, bad), vehicle (nearly perfect), and accident (avoidable, preventable) proved remarkably successful throughout the early 1900s. Eastman offers a few examples of how this carefully crafted understanding was readily accepted and circulated by the popular press: "An editorial in Collier's in 1925 stated, 'Automobiles are now nearly fool-proof. Streets are not and some drivers are fools,' and the magazine repeated six years later, 'Automobiles are built for safety but we throw caution to the wind and reap the harvest of recklessness.' Another journal summed it up less colorfully a year later when it reported: 'Death and injury are to be attributed in the main to the predominance of inefficient but otherwise well meaning drivers.'"[22]

The single most widely read and culturally influential article on the subject of automobile accidents to appear during this period was J. C. Furnas's "—And Sudden Death." Originally published in Reader's Digest in August 1935, the article describes the car-crash experience—and especially its traumatic effects on the human body—in truly excruciating detail. "Like the gruesome spectacle of a bad automobile accident itself," Furnas writes by way of preface, "the realistic details of this article will nauseate some readers. Those who find themselves thus affected at the outset are cautioned against reading the article in its entirety, since there is no letdown in the author's outspoken treatment of sickening facts."[23]

No letdown, indeed. Even today Furnas's article is striking for its lurid language, gory imagery, and macabre tone. Throughout, the motorcar comes across as something like a medieval torture instrument, one equipped with a "lethal array of gleaming metal knobs and edges and glass."[24] "Lethal" because, in a high-speed accident, "every surface and angle of the car's interior immediately becomes a battering, tearing projectile, aimed squarely at you—inescapable. . . . It's like going over Niagara Falls in a steel barrel full of railroad spikes."[25] Paragraph after paragraph is replete with moaning voices, mutilated bodies, mangled corpses. In relentless succession, the reader is told of Z-twisted legs and blood-dripping eyes; "bones protruding through flesh in compound fractures" and "dark red, oozing surfaces where clothes and skin were flayed off"; brains pierced by wooden fragments and skulls shattered by dashboards; abdomens impaled by steering columns and bodies decapitated by windshields; internal injuries and hemorrhages, smashed hips and knees, cracked ribs and collarbones, broken pelvises and spines.[26] "As Furnas tries to combat the abstracting effect of statistical reports by asserting the individuality of specific crash victims," Karen Beckman observes, "bodies repeatedly lose their limits, turn inside out, and merge into each other."[27]

"—And Sudden Death" immediately aroused a great deal of public interest and concern, and its moral-pedagogical potential was not lost on civic authorities. In the months following its publication, millions of reprints were ordered and distributed by schools, churches, social clubs, safety organizations, police departments, and traffic courts. Additionally, the article was republished in some two thousand newspapers and magazines. Eastman notes that "the total printed circulation of '—And Sudden Death' in the last few months of 1935 was estimated at over 35,000,000 copies."[28]

At the start of the article, Furnas stressed that he wanted to move beyond the impersonal abstractions of injury and fatality statistics in order to bring "the pain and horror" of motor-vehicle accidents "closer home."[29] But to what end? Why did Furnas think it necessary to bring home such pain and horror to the readership of *Reader's Digest* and, by extension, presumably, to white, conservative, middle-class America? What did he hope to accomplish in doing so? For that matter, why did the magazine's editor, DeWitt Wallace, commission Furnas to write the piece in first place? After all, *Reader's Digest* was not exactly known for its sensationalist stories, much less for its morbid fascinations.

Furnas explained that his purpose was to jar "the motorist into a realization of the appalling risks of motoring."[30] For such a realization to be effective, he added, it must be more than fleeting: "What is needed is a vivid and *sustained* realization that every time you step on the throttle, death gets in beside you, hopefully waiting for his chance."[31] Wallace, for his part, "immediately realized that if more people knew what an accident was really like it might bring some of the reckless drivers and speed maniacs to their senses."[32] Hence the editor's addendum, which appeared on the last page of the article: "Convinced that widespread reading of this article will help curb reckless driving, reprints in leaflet form are offered at cost. . . . To business men's organizations, women's clubs, churches, schools, automobile clubs, or other groups interested in public welfare, we suggest the idea of distributing these reprints broadcast. The cover of the leaflet provides space for any message or announcement of your own that you may wish to add."[33]

Together, Furnas's and Wallace's stated intentions suggest the extent to which the figure of the reckless driver, causer of accidents, had become commonplace by the 1930s. One might even say that the reckless driver had become a cultural archetype of sorts, a new kind of outlaw or antihero for a new kind of society and structure of feeling, a troublemaker all the more treacherous on account of his or her ordinariness. Furnas's and Wallace's comments suggest, as well, the extent to which driver education, the third of the so-called Three E's, was thought to be an effective, if partial, remedy for recklessness.

"—And Sudden Death" was commissioned, written, and edited with the express purpose of reforming reckless drivers and dissuading those who, in a moment of rashness, might be tempted to join their ranks. Accordingly, it frequently addressed the reader, assumed to be a motorist, in the second person. From start to finish, the article's tone remains urgent, admonitory, didactic. If you do not want to end up like these victims, it warns, then you must drive carefully and conscientiously at all times. When you mix "gasoline with speed and bad judgment," this is the sudden death or the protracted agony that awaits you.[34] "If you customarily pass without clear vision a long way ahead, make sure that every member of the party carries identification papers—it's difficult to identify a body with its whole face bashed in or torn off."[35]

And so on. Nowhere in the piece is the automobile itself indicted or even held under suspicion, only its imprudent operator. Drivers exercise "bad judgment," Furnas insists, not the designers, engineers, manufacturers, or marketers of the vehicles that drivers use—or, rather, that they *mis*use. The way to prevent mishaps, therefore, is to educate motorists as to the high risks and horrible realities of behaving badly behind the wheel. Wallace and Furnas hoped and believed that "—And Sudden Death," with its shocking, sickening images of human carnage, would serve as an object lesson in just such an education. And so did the schools, churches, social clubs, safety organizations, police departments, and traffic courts that bulk-ordered and mass-distributed reprints of the article.

The particular complex of driver, vehicle, and accident I have been examining operated as a powerful social, ethical, and institutional discourse in the United States during the first half of the twentieth century. It worked to fix certain meanings, cultivate certain attitudes, inculcate certain behaviors, and assign certain obligations in relation to "safe" automobility. It was integrally involved in the normalization and symbolic regulation of the emergent car culture, and its ascendance marked a distinctive moment the history of mechanized travel and accelerated mobility. Crucially, each of its terms—driver, vehicle, accident—carried its own normative assumptions. These were: the driver is an autonomous subject who freely chooses to act either responsibly or recklessly and who, because he or she is essentially rational, can be trained to choose the former; the vehicle is an intrinsically safe and reliable technology without agency of its own; and the accident is an entirely unnecessary occurrence, typically resulting from a simple and singular cause: the reckless driver, the speed maniac.[36] Here, then, is a macro-discourse of safe automobility composed of three mutually informing micro-discourses: a discourse of *subjectivity* linked to a discourse of *technology* linked to a discourse of *accidentality*. Let us consider these micro-discourses in turn.

SUBJECTIVITY

The driver is an intentional agent, a rational being, and a sovereign individual. He or she assesses, decides, and acts accordingly, for better or for worse. His or her behavior is self-determined and self-directed, and it holds both material consequences and societal implications. Because driving is an expression of individual volition, reckless driving is amenable to moral reform: it can be corrected through a combination of instruction (the third E, education) and punishment (the second E, enforcement). This ethico-juridical formulation rehearses long-standing, deeply entrenched liberal-humanist ideals and values—namely, in Katherine Hayles's words, "a coherent, rational self, the right of that self to autonomy and freedom, and a sense of agency linked with a belief in enlightened self-interest."[37] Insofar as he or she was presumed to be endowed with intentionality, reason, self-sovereignty, and a seemingly infinite capacity for self-improvement (moral perfectibility), the driving subject of the early twentieth century recalled and recontextualized the liberal-humanist subject of the Enlightenment. Also, and more immediately, this speed-appreciating subject recycled distinctly American fantasies of freedom and personal mobility (themselves partly rooted in liberal humanism's valorization of self-movement and self-regulation) as well as of democratic individualism and decentralized control.[38]

TECHNOLOGY

In legal parlance, the term *dangerous instrumentality* denotes a category of objects considered inherently harmful or hazardous, such as firearms, explosives, and ferocious animals. Under the law, "the owner of a dangerous instrumentality [has] a special obligation to keep it with care."[39] Sarah Lochlann Jain has shown how the American judicial system repeatedly refused to recognize the motor vehicle as a dangerous instrumentality in the early decades of automobility. Instead, the courts most often regarded it as an ordinary object, a consumer product like any other, a thing that "merely extend[s] a driver's will."[40] Judges, in their rulings, literally legitimated the conception of the car as "a nearly perfect mechanism"—a conception that, as we have seen, informed the reasoning and inflected the rhetoric of the popular press, the automobile industry, the highway-safety movement, and civic authorities. And so, dizzily, the discourse spun round: if the automobile is not inherently harmful or hazardous, then it is not legally a dangerous instrumentality; and if it is not legally a dangerous instrumentality, then it is not inherently harmful or hazardous; and if it is not inherently harmful or hazardous—if it is not like a gun or a bomb or an attack dog—then its operator is under no "special obligation to keep it with care."

The courts' take on the rights and responsibilities of automobility was allied with a broader, well-established rationality of technology. Already part of the cultural "common sense," this rationality would have predisposed the public to accept and adopt, rather than to contest or refuse, the perspective on safe automobility advocated or propagated by the aforementioned industries and institutions. Just as the eighteenth-century liberal-humanist subject prepared the ground for the twentieth-century driving subject, so a prevailing technological rationality—a particular way of thinking about artifacts as useful and, at the same time, of relating to useful artifacts—provided a notional and rhetorical framework for apprehending the true essence, as it were, of the automobile.

This technological rationality executes two basic moves. First, it reduces a sophisticated technology to the status of a simple tool; second, it presupposes the social, ethical, and political neutralities of that tool. In this way, it exemplifies what Andrew Feenberg calls an "instrumental theory" of technology. Instrumental theories conceive of technology as a means "indifferent to the variety of ends it can be employed to achieve," a mechanism "without valuative content of its own."[41] Technologies are merely things—benign, unbiased, devoid of agency—that stand "ready to serve the purposes of their users."[42] Langdon Winner makes a similar critical point:

> According to conventional views, the human relationship to technical things is too obvious to merit serious reflection. . . . Once things have been made, we interact with them on occasion to achieve specific purposes. One picks up a tool, uses it, and puts it down. One picks up a telephone, talks on it, and then does not use it for a time. A person gets on an airplane, flies from point A to point B, and then gets off. The proper

interpretation of the meaning of technology in the mode of use seems to be nothing more complicated than an occasional, limited, and nonproblematic interaction.[43]

Accompanying this conventional "meaning of technology in the mode of use" is a conventional morality of technology: "Tools can be 'used well or poorly' and for 'good or bad purposes'; I can use my knife to slice a loaf of bread or to stab the next person that walks by. Because technological objects and processes have a promiscuous utility, they are taken to be fundamentally neutral as regards their moral standing."[44]

Feenberg's and Winner's critiques help us to see how an ideologically dominant instrumentalist rationality of technology conditioned the possibility of the automobile's widespread social, ethical, and political neutralization—of its tendentious reduction to the status of an ordinary object—during the first half of the twentieth century. As "the most widely accepted view of technology," instrumentalist rationality lent credence to the notion that cars were tools to be used for good or for ill, that they were machines empty of ethical import, no different in principle from electric toasters, automatic dishwashers, or any other consumer appliances.[45] At the same time, it worked to deflect attention from what Ralph Nader famously called "the designed-in dangers of the American automobile," obscuring the social agencies of the vehicle engineer, the vehicle manufacturer, and the vehicle itself.[46] The cultural predominance of technological instrumentalism thereby abetted and facilitated the popular acceptance of the idea that cars "don't kill people, people kill people," including themselves.

ACCIDENTALITY

Remember what Eastman said about the National Safety Council: "The rationale for the Council's approach to accidents was based on the assumption that all accidents had causes; as one author put it: 'Human misbehavior, human frailty, human ignorance, human laziness all cause accidents. Remove them and accidents cease.'" This assertion encapsulates the discursive construction of the accident in early twentieth-century car culture. According to the logic of this construction, road mishaps are caused by human actions; such actions are correctable because humans are perfectible; such perfectibility means that—here is the key point—road mishaps are *eliminable*. As Jain observes, "The obvious corollary of assuming that the car is an ordinary object was that accidents would be understood as the result of driver negligence. Perfect driving was assumed to be humanly possible and legally obligatory."[47] Smashups were direct effects of human mistakes, ignorant or imprudent deviations from a code of behavior that should, and could, be followed flawlessly. Operator errors were unnecessary; "remove them and accidents cease." This fantasy of the perfect driver—and, by extension, of a nation of perfect drivers—rearticulated a mythos of human perfectibility in the context of an emergent culture of automobility. It imagined a world in which all reckless drivers and speed maniacs would be rehabilitated and in which all traffic accidents would be obviated. It advanced a logic of simple subtraction—accident-causing errors were to be eliminated one by one, through the implementation of the Three E's—in order to arrive at a smoothly functioning, totally safe road-transport system.

The Turn to Crash Injuries and Crashworthiness

The driver-negligence paradigm did not go entirely unchallenged during the period of its predominance, from roughly 1900 to the 1940s. Beginning in the 1930s, a few individuals, most of them medical professionals, began to seriously question the wisdom of concentrating on driver attitudes and behavior. Instead, they focused on the damage done to bodies, on the mechanical forces and processes involved in the production of that damage. They wanted to know why crash injuries occurred and what could be done to reduce their number. Rather than worrying about the psychic interior of the driver (his or her morals, motives, mind), they worried about the physical interior of the vehicle (its surfaces, edges, protuberances). Rather than seeking to eliminate mishaps by reforming the motorist, they sought to mitigate the harmful effects of mishaps by rethinking the design of the machine. Rather than acceding to the fantasy of driver perfectibility, they accepted the reality of the accident's inevitability.

Claire L. Straith, a plastic surgeon from Detroit, was a pioneering figure in the effort to radically recast the problem of automobile accidents during the 1930s.[48] Straith spent countless hours and much of his career reconstructing the bodies, especially the faces, of car-crash victims. "I have seen the torn and mutilated victims of crashes for nearly four decades," he declared in 1957. "At least we have the chance to help the victims, the ones lucky enough to survive, to return to normal appearances."[49]

Straith customized his own car for safety in the early 1930s, installing seatbelts and affixing to the dashboard crash padding of his own design and manufacture.[50] In 1934, he initiated what would become an ongoing dialogue with automobile-industry executives and engineers, urging them to remake the vehicle interior so as to lessen the frequency and severity of accident-related injuries. A year later, Straith met with Walter P. Chrysler, founder and head of Chrysler Corporation, who, along with the company's chief engineer, responded favorably to his ideas and suggestions. Debuted in the fall of the following year, the 1937 Dodge incorporated several Straith-inspired features, including inward-curved door handles, nearly flush window regulators and a completely encased windshield regulator (it was not unusual in those days for windshields to open for ventilation), non-protruding instrument-panel buttons, a raised dashboard to reduce knee injuries, and a rubber-padded front-seat back to protect rear-seat passengers.

As Eastman points out, other automakers such as Studebaker Corporation "adopted some of Chrysler's innovations while public concern over safety remained high in the later 1930s"—concern provoked, in no small part, by Furnas's *Reader's Digest* article.[51] Nevertheless, in the absence of either government regulation or a reliable gauge of success, such innovations were soon sacrificed on the altar of the "annual model change."

> The annual model change required new styling of the interior, as well as the exterior, and when the new models appeared after the war, the "safety smooth" dashboards, recessed door handles, and other small improvements had been replaced by new and different, but usually less safe, designs. A promising beginning of automobile design for crash protection had been brought to a halt by the demands of the annual model change. Chrysler safety engineer Roy Haeusler later reported that there was no way to evaluate the effectiveness of his company's instrument panel design, and, thus, no way to justify its retention—so it was changed for styling purposes.[52]

By the late 1940s, Straith was not the only physician researching car-crash injuries and recommending ways to redesign the vehicle interior for occupant safety. In June 1948, Fletcher D. Woodward, an otolaryngologist (ear, nose, and throat specialist) from the University of Virginia Hospital, delivered an address titled "Medical Criticism of Modern Automotive Engineering" at the American Medical Association conference in Chicago. Published later that year in the *Journal of the American Medical Association*, the address defended the medical profession's increasing interest in the relationship between automobile design and accident injury. Woodward proposed that automotive engineers take their cue from aeronautical engineers, who, in recent years, had minimized crash injuries by modifying cockpit instrumentation in light of studies conducted in the field of aviation medicine. He went on to specify a number of ways in which cars might be built for safety rather than for speed and style. Most interesting for our purposes, however, was Woodward's explicit attempt to reframe questions of agency, accountability, and accidentality:

> In spite of the success of [highway-safety] campaigns, the automobile still remains a lethal and crippling agent, and since there appears to be little likelihood of accomplishing radical changes in human nature in general and exuberant youth in particular, it would seem the part of wisdom to shift the emphasis, for the moment at least, to desirable alterations in the machine itself, rather than to place all emphasis on attempts to bludgeon "old Adam" into safer driving practices.
>
> Our highways are becoming increasingly laden with cars driven by average persons, and it appears inevitable that these machines will continue to collide, pass on turns, fail to observe stops signs, leave the road at high speeds and afflict mankind much as they have in the past.[53]

This is a fairly astonishing statement for its day. In a few carefully worded sentences, Woodward repudiates five decades of thought and practice in the area of automotive safety. He pays lip service to the highway-safety movement, to be sure, but it is clear that he considers the movement a failure; indeed, not only the nature but the very fact of his proposal presumes that failure. A newspaper reporter in 1902 wrote, "The accident which happens to the automobile is seldom due to the machine itself, but almost wholly to the loss of control or presence of mind of the operator." Forty-six years later, Woodward aims to reverse this line of reasoning: awkward or absent-minded operators are now "average persons" victimized by automobiles that "continue to collide, pass on turns, fail to observe stops signs, leave the road at high speeds." The seat of agency, once singular and synonymous with the driver's psyche, is now doubled and dispersed: the machine, no less than the man or woman, exerts a kind of influence, exercises a kind of power, *acts*. Human nature cannot be changed by the bludgeons of the Three E's, no matter how hard or how often they strike. The "lethal and crippling" nature of the automobile, on the other hand, can be changed with comparative ease, using nothing more than the ideas and implements of the modern engineer. As the total number of cars on the nation's roads climbs with each passing day, it becomes increasingly difficult to sincerely sustain the dream of accident eliminability. Instead, the accident "appears inevitable."

Woodward's address to members of the American Medical Association in 1948 was clearly informed by Straith's automobile-design innovations.[54] But it probably owed an even bigger debt to the work of a self-trained engineer and pathologist from New York named Hugh De Haven. "Although De Haven is now almost unknown," writes Karen Beckman in *Crash: Cinema and the Politics of Speed and Stasis*, "we live today more than ever in the wake of the paradigm of human safety he developed."[55]

De Haven's interest in crash injuries began in 1917 while he was a cadet in the Canadian Royal Flying Corps, which he joined after failing to meet U.S. Army Air Corps requirements. One day during a flight-training exercise, his Curtiss JN-4 "Jenny" collided with another plane and crashed to the ground. De Haven suffered serious injuries, including two broken legs, assorted bruises and lacerations, and a ruptured liver, pancreas, and gallbladder. During his six-month in-hospital convalescence, he became preoccupied with what he would later call "the mechanics of injury and safety design."[56] His colleagues were convinced that he was "lucky" to have survived the accident, that his victory over death was a "miracle." But, as Howard Hasbrook noted in an article for *Clinical Orthopaedics* in 1956, "De Haven felt that the intactness of [his plane's] cockpit structure was the answer; thus was born the first concept of 'crashworthiness.'"[57]

De Haven continued to pursue the concept of crashworthiness after leaving the hospital. "When De Haven returned to active duty, he was given . . . the job of rushing to local airplane crashes which happened frequently around the training field, and he had an opportunity to study crashes and related injuries in great detail and numbers."[58] Recalling these formative experiences, De Haven later wrote, "Observations made at that time, during investigation of air crashes, gave strong indication that many of the traumatic results of aircraft and automobile accidents could be avoided. Structures and objects, by placement and design, created an inevitable expectance of injury in even minor accidents."[59] The safety recommendations De Haven made on the basis of these observations, however, tended to be either written off or ridiculed. Pilots scoffed at the suggestion that cockpits and cabin structures be reengineered for crashworthiness. (In the military aviator's macho code of honor, such a concern was tantamount to cowardice.) Other professionals, too, dismissed or mocked him. Nader writes, "De Haven was turned away repeatedly by government and university people who called him a 'crackpot.' In the twenties and thirties, he recalls, 'The saying used to be, "If you want to be safe, don't fly."'"[60] Airlines and aircraft manufacturers turned De Haven away as well.

It was not until 1942 that De Haven's research came into public view, with the publication, in the American Medical Association's journal *War Medicine*, of "Mechanical Analysis of Survival in Falls from Heights of Fifty to One Hundred and Fifty Feet." In the paper, De Haven assembled and analyzed eight documented cases of human free fall.[61] Most were suicide attempts, some were accidents, but in each case the faller survived, as if miraculously, often sustaining only minor injuries. De Haven's stated objective in studying these strange occurrences was, first, "to establish a working knowledge of the force and tolerance limits of the body" and, second, to facilitate the application of that knowledge to airplane and automobile design.[62] By casting a clinical eye on instances in which ordinary individuals walked away (literally, in at least one case) from downward plunges that, statistically speaking, should have killed them, De Haven revealed the human body to be far more resilient to extreme force impact than previously imagined. More to the point, his research proved that falls from great heights and like phenomena were survivable, provided that impact pressures were cushioned and dispersed, or, as De Haven put it, "distributed in distance (time) and area (space)."[63] "Extreme force within limits can be harmless to the body," he wrote. It was the "structural environment," too stiff and too sharp, that was "the dominant cause of injury."[64] Seeing in these findings major implications for transportation safety, and voicing ideas that Claire Straith had custom-built into his own car a few years earlier, De Haven called for airplane and automobile interiors to be structurally engineered for the purpose of reducing the severity of crash injuries and the frequency of crash fatalities.

Scientifically grounded and empirically substantiated, the *War Medicine* article helped De Haven overcome the "crackpot" designation. More important, though, it enabled him to obtain moral and material support from the Safety Bureau of the Civil Aeronautic Board, the National Research Council's Committee on Aviation Medicine, and the Office of Scientific Research and Development. A grant from the latter led to the creation in 1942 of a research center "dedicated," in Beckman's phrase, "to rendering Americans in motion 'deathproof'": the Crash Injury Research Project at Cornell University Medical College in New York City.[65]

At Cornell, De Haven and his Crash Injury Research group developed a novel and sophisticated system for accumulating and analyzing data pertaining to survivable light-plane accidents and related injuries. Specially prepared forms were distributed to "highly motivated state police and state aviation groups," rather than to the Civil Aeronautics Administration, as might be expected (and was later done), since at that time "the Civil Aeronautics Act did not provide for investigation into the causes of injuries—only the causes of accidents."[66] In an article for the journal *Public Safety*, De Haven explained that project participants would use the forms to record "careful estimates . . . of impact speeds and stopping distances which, with photographs of the general wreckage, cabin interior, seats, instrument panels, control wheels and other details, allow judgment of accident force and severity. By analyzing, cross referencing, and filing data, it soon was possible to identify specific structures and conditions which repeatedly caused similar injuries in similar accidents."[67] This accident-reporting system represented an important practical and philosophical reorientation in air-crash investigation in the United States. Prior to the Crash Injury Research Project, accident investigators were concerned to ascertain the cause of the crash, a forensic concern that has not diminished to this day. But the new system—and the new, medicalized way of thinking it embodied—required that at least some accident investigators be attentive to the causes of crash injury and fatality.[68] In this respect, the new system hinted at the inevitability and the ineliminability of the accident. It also prescribed an *autoptic* way of seeing—including an autoptic way of *photographically* seeing—post-crash "structures and conditions" in relation to human anatomies. It specified a new object for the forensic gaze.

De Haven knew that alleviating the causes of human injury meant convincing manufacturers to make their machines less dangerous to use and occupy—not less dangerous to use and occupy during the countless hours of "normal operation," but rather during the split second of catastrophe, during that miniscule interval in which there is no operation as such (and no "operator," either). It meant getting manufacturers to rethink their responsibility—in fact, it meant getting them to *redesign* their responsibility, to literally build an obligation to safety into their products. In a 1950 technical report summarizing a study jointly sponsored by the U.S. Navy, the U.S. Air Force, and the Civil Aeronautics Administration, De Haven wrote, "All details of these [accident-investigation] reports are recorded on punch cards and analyzed for characteristic 'patterns of injury' resulting from specific structures. Dangerous structures are called to the attention of the manufacturer and recommendations are made for redesign to minimize known dangers. The information obtained has been of assistance to manufacturers in considering the overall configuration and arrangement of structure in new planes as well as in the redesign of seat backs, instrument panels, control wheels, seats, safety belts, flooring, and other details."[69]

By the early 1950s, the Crash Injury Research Project's scope had expanded to include the study of automobile injuries, and by the middle of the decade a reporting system similar to the one used for light-plane accidents had been implemented in eighteen states.[70] In several of his published writings on the subject, De Haven, like Fletcher Woodward, made explicit his intention to shine the national highway-safety spotlight, not on poorly disciplined drivers and accident prevention, as had been done for decades (with limited results), but on poorly designed vehicles and injury prevention: "Despite nationwide accident prevention programs, the steadily increasing use of airplanes and automobiles is resulting in a greater number of accidents each year. Consequently there is an increasing need of safety engineering and design to offset the danger of injury in crashes."[71]

During the war, De Haven maintained that "evidence of the extreme limits at which the body can tolerate force cannot be obtained in laboratory tests for obvious reasons."[72] After the war, those reasons no longer seemed so obvious, at least not to John Paul Stapp, whose legendary postwar rocket-sled experiments proved that the human body could endure at least 35 g-forces of deceleration in the forward-facing seated position, nearly twice as many as scientists had considered the survivable limit.[73] Stapp concluded from this, in the spirit of De Haven, that airplane structures and objects (cockpits, cabins, seats, harnesses) should be designed to handle deceleration forces of commensurate magnitude. In coming years, he would say the same of automobile structures and objects.

During his tenure at the Wright Air Development Center's Aeromedical Laboratory in Dayton, Ohio, Stapp field-tested liquid-oxygen emergency breathing systems for military pilots and recommended ways to prevent high-altitude bends, chokes, gas pains, and dehydration.[74] In May 1947, Stapp was moved to Edwards Air Force Base (then known as Muroc Army Air Field) in California's Mojave Desert, where he organized an aeromedical field laboratory like the one at Wright Field (but with far fewer resources) and served as project officer—and test subject—for the first rocket-sled research program. This program was designed to study the limits of human tolerance to rapid deceleration as well as the strength of airplane seats and harnesses under simulated crash conditions.

On 10 December 1947, after carrying out more than thirty runs using anesthetized chimpanzees or anthropomorphic dummies, Stapp strapped himself into the Northrop Aircraft "Gee Whiz" Decelerator Sled. When the speeding machine came to a sudden stop a few seconds later, the first rocket-sled experiment to use a live human volunteer was over, and a new kind of transportation-accident research had begun. Stapp had, in effect, taken De Haven's free-fall studies—specifically their concern with the survival limits of rapid deceleration and extreme force impact—and turned them into technically complex, scientifically controlled field tests. Stapp not only gathered crash-injury information; he generated it.

By May of the following year, Stapp had ridden the rocket sled sixteen times and had been subjected to deceleration forces, or "Gs," equivalent to thirty-five

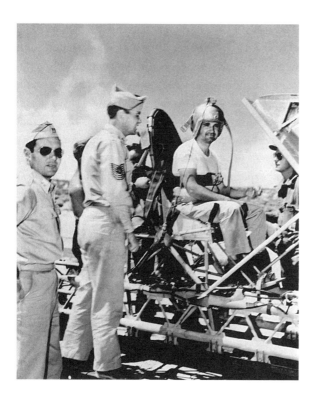

John Paul Stapp aboard rocket sled, circa 1947. Courtesy of Edwards Air Force Base History Office.

times the pull of gravity. "The men at the mahogany desks thought that the human body would never take more than 18 gs," Stapp told *Time* magazine in 1955. "Here we were, taking double that—with no sweat."[75] In June 1951, Stapp completed his work at Edwards, where he survived a total of twenty-six trips down the test track, sustaining only minor injuries. His dangerous, unorthodox research had demonstrated that the human body, if properly positioned and restrained, could withstand at least forty-five Gs at 500 Gs per second rate of onset.[76] And if forces this extreme were not, in and of themselves, lethal, as had been shown to be the case, then Stapp's hero, Hugh De Haven, was quite right in saying that high-speed, heavy-impact "crackups" were survivable, given the right "structural environment"—that is, a structural environment engineered for crashworthiness.

In April 1953, Stapp assumed command of the Holloman Air Force Base's Aeromedical Field Laboratory in New Mexico, where he continued his work on the effects of mechanical force on living tissue. His personally designed and directed research project, Biophysics of Abrupt Deceleration, investigated the problem of high-speed, high-altitude escape from aircraft, or "escape

Tests and Split Seconds 163

physiology."[77] Whereas Edwards's "Gee Whiz" only had the capacity to study the effects of rapid deceleration, Holloman's high-speed track, originally built in 1949 as a rail launcher for the Snark missile, had the capacity to study the effects of rapid deceleration in combination with those of windblast and tumbling. It was 1,550 feet longer than the Edwards rocket sled, and it featured an advanced system of water brakes (rather than mechanical-friction brakes) that "permitted both high deceleration forces and a wide range of duration and rate of onset."[78]

For Stapp's latest round of experiments, Northrop Aircraft created a rocket sled called Sonic Wind No. 1. Practice runs began on 23 November 1953, and the first run with a living subject—a chimpanzee—took place on 28 January 1954. (Chimps were the animal of choice for rocket-sled research, but hogs and black bears were "sacrificed" in related experiments at Holloman.) With authorization from Air Research and Development Command Headquarters, Stapp took his first ride in Sonic Wind No. 1 on 19 March of that year, reaching a peak velocity of 615 feet per second. To explore the effects of abrupt windblast, a second human experiment was conducted seven months later, with Stapp again serving as test subject.

Stapp's final, and most celebrated, rocket-sled experiment took place on 10 December 1954, seven years to the day after he first climbed aboard the "Gee Whiz." On that late-autumn morning in 1954, Sonic Wind No. 1, with Stapp onboard, reached a record-breaking 632 miles per hour in five seconds—faster than the Lockheed T-33 observation aircraft flying overhead—then slammed to a halt in 1.4 seconds. The experiment subjected Stapp's body to forces in excess of forty-six times the pull of gravity—the most any human being has ever deliberately endured—and it temporarily blinded him. It also made him a national hero and a media icon. In the wake of his record-breaking ride, Stapp made several television appearances, including one on Ralph Edwards's *This Is Your Life*. Casting the scientist as superhero, *Collier's* dubbed Stapp the "Fastest Man on Earth" in a June 1954 cover story. *Time* did its own "Fastest Man on Earth" cover story on Stapp in September of the following year, and Twentieth Century Fox's *On the Threshold of Space*, a romanticized Hollywood B-movie partly based on Stapp's rocket-sled research, appeared the year after that.

As early as 1948, Stapp was aware that his experiments, undertaken in the name of aviation medicine, held profound implications for automotive safety, and over the next few years he became increasingly interested in the issue of automotive crashworthiness. Soon he was saying the sort of thing that could have come out of the mouth of Straith, Woodward, or De Haven: "Interiors of vehicles should be delethalized in order to increase their crash protective

John Paul Stapp on the cover of *Time* (September 1955).

characteristics with specific attention given to any object which could cause injury if impacted by the occupant."[79]

In 1953, he proposed that the Aeromedical Field Laboratory carry out a series of controlled tests in order to measure actual car-crash forces and "to establish criteria for modifications and specifications for vehicles, personnel restraints and . . . regulations for automotive safety."[80] Although some of his superiors did not approve of automotive research being conducted under the auspices of an aeromedical research institution, Stapp was able to persuade the decision-makers that car-crash injuries and fatalities were something the military should take seriously. The statistics backed him up. At the time, automobile accidents ranked second as a cause of death and first as a cause of hospitalization among Air Force personnel, and first as a cause of death among Army personnel.[81] In 1954, the Air Force lost some 700 men in plane crashes and 628 in car crashes.[82] Such losses, Stapp insisted, constituted "a needless waste of manpower to the defense effort. . . . Driver's education will help somewhat to prevent accidents but concrete steps should be taken to protect the vehicle occupant when the accident occurs."[83] This last statement neatly epitomizes the conceptual and discursive turn to crash injuries and crashworthiness we have been considering: driver's education is dutifully mentioned but only faintly recommended ("somewhat"); the primary imperative is to protect the vehicle occupant, not to prevent the accident; and the occurrence of that accident is a regarded as a matter of "when," not "if."

Full-scale crash-testing began in the spring of 1955. An uninstrumented trial run, using a 1945 Dodge Weapons Carrier and a pair of lap-belted dummies, took place in March 1955. Two months later, in front of an audience of invited guests from the ranks of industry, government, and academia, a fully instrumented crash test was staged, formally inaugurating the Aeromedical Field Laboratory's new Automotive Crash Forces project. (This turned out to be the first of three Annual Automotive Crash Research and Field Demonstration conferences held at Holloman.) Various kinds of crash test, along with various kinds of smaller-scale force-impact experiment, were conducted over the next three years, using anesthetized animals, anthropomorphic dummies, and even human volunteers (not Stapp, however). Air Force salvage vehicles were rolled over while in motion, crashed into fixed barriers, and crashed into other vehicles. In some tests, vehicles were brought to an abrupt halt by means of metal cables attached to their frames. Hogs were restrained in specially designed "swing seats" and released like raised pendulums into steering wheels and columns. Other specially designed machines were employed as well, including a crash-restraint demonstrator, originally built for aircraft and informally known as "The Bopper," and a short, two-rail deceleration device called "The Daisy Track."

De Haven's and Stapp's medically oriented investigations helped to shift the institutional focus of — and to change the national conversation about — automotive safety in the United States. Before the 1940s, the overriding question had been "How can accidents be prevented?" and the answer, "By reforming the reckless driver." (Implicit in this formulation is the notion that mishaps are eliminable because the motorist is perfectible.) By the end of the 1940s, however, the overriding question had become "How can crash injuries and fatalities be reduced?" and the answer, "By reengineering the automobile." (Implicit in this formulation is the notion that mishaps are inevitable but the motorcar is improvable.) For De Haven and Stapp, the primary imperative was not to retrain the mind that moved the machine that moved the body, but rather to rebuild the machine in line with the body's newfound robustness, bypassing the messy problem of the mind altogether. Just as human anatomy had been shown to be that much stronger, so technology would have to be that much safer. The trouble was not that the transported organism was naturally defenseless; it was that the transport mechanism was unnecessarily dangerous. To use the professional jargon of the period, cars needed to be *delethalized*, engineered for *crashworthiness*, designed according to empirically established standards of *survivability*.

Whether studying injury phenomena associated with human free falls or with light-plane accidents, De Haven arrived at his findings by analyzing and interpreting data gathered from the field. Stapp, on the other hand, arrived at his findings by analyzing and interpreting data generated under laboratory-like conditions. The former looked for evidence pertaining to real-world impacts and crashes that had already happened, whereas the latter deliberately created impacts and crash simulations to be looked at—evidenced—as they were happening. This methodological distinction marked a turning point in the history of transportation-safety research. For Stapp, accidents and the injuries that accompanied them were not only something that should be forensically investigated after the fact; they were something that could be forensically *prepared* through artificial staging and scientific witnessing. And no scientific witness was considered more reliable and revealing—more forensically useful and valuable—during this period than the high-speed film camera.

The Development of Scientific Crash-Testing

Holloman's Aeromedical Field Laboratory was not the only organization in the United States during the 1950s carrying out scientifically controlled car-crash and force-impact tests. On the contrary, a number of public and private institutions, most of them research universities, as well as a few automobile manufacturers, most notably Ford Motor Company, were doing much the same, and their experimental findings and automotive-engineering recommendations quickly circulated among interested individuals and organizations. Practically everything about these tests was unprecedented: their rationale, methods, and objectives; their technical, scientific, and administrative sophistication; their scale, variety, and sheer quantity. Never before had motor-vehicle crash-test research been carried out with such rigor and attention to detail, with so much high-tech equipment and so many labor-intensive hours, in and across so many industrial and institutional sites. Throughout the 1950s and 1960s, such research extended—in spectacularly destructive fashion—the quest for knowledge about the causes of crash injuries and about the benefits (and limits) of crashworthiness.

Some of the most significant crash-test research during the postwar period took place at Cornell University. While the Crash Injury Research group at the Medical College in New York City was busy gathering, cataloguing, and analyzing data pertaining to actual road mishaps, researchers at the Cornell Aeronautical Laboratory in Buffalo were busy creating, controlling, and learning from their own "accidents." As Edward R. Dye, Head of the Laboratory's Industrial Division, explained in 1955, "Our research at the Laboratory in Buffalo is aimed toward the collection of engineering information, by mathematical analysis and/or applied research techniques, on the factors which cause injury to the human and [toward] the use of the information so collected to design ways of increasing the safety in the mechanical automobile device."[84]

Cornell Aeronautical Laboratory's earliest impact experiments, dating from the late 1940s, saw hen eggs standing in for human heads.[85] After the Crash Injury Research group discovered that approximately three-quarters of all deaths in light-plane crashes were due to head injuries, the laboratory began researching ways to protect the vulnerable pates of aircraft pilots and passengers. Reasoning that both eggs and heads are "hard shells surrounding a semi fluid and are, in general, ellipsoidal in shape," researchers dropped eggs from various heights onto various energy-absorbing materials and recorded the results.[86] Additionally, they developed a small swing-like contraption in which an egg was placed and then crashed against a concrete column. In another set of experiments, sponsored by the Medical Science Division of the Office of Naval Research, gelatin-filled human skulls sealed with cellophane tape were dropped from various heights onto hard, flat surfaces (and, later, onto differently shaped surfaces) in order to ascertain fracture thresholds and rates and heights of rebound. In related experiments, plastic head-forms were catapulted at speeds of up to forty miles an hour into cockpit instrument panels.[87] Each of these tests, significantly, was studied with the aid of a high-speed motion-picture camera, a medium that was central to postwar crash-testing.

In 1953, the laboratory, with sponsorship from the Liberty Mutual Insurance Company of Boston, undertook an elaborate automotive crash-safety research project. The project had three basic objectives: to measure the kinematic motions of belted and unbelted occupants during the crash-deceleration interval; to find ways and means of reducing the injurious effects of those motions; and to find ways and means of reducing the monetary costs of crash-caused vehicular damage.[88] To achieve the first objective, intricate "crash-snubbing" tests were staged, using a 1950 two-door, five-passenger Ford sedan and an unmatched pair of specially designed human surrogates: "Thick Man," modeled on an adult male, and "Half Pint," modeled on a six-year-old child.[89] ("Thin Man," a cruder sheet-metal dummy, was used extensively in earlier, aircraft-related experiments at the laboratory.) According to Dye, the dummies were carefully developed by the laboratory "to behave, in these violent adventures, exactly as living passengers would."[90] The dummies' kinematic motions—their precise positions and airborne trajectories during the deceleration interval—were captured and analyzed by means of high-speed motion-picture cameras. "Because the action is so fast," Dye explained in an article for *Woman's Day* in 1954, "we collect our data with special electronic instruments and high-speed movie cameras; later we can study the accident in slow motion. This lets us see exactly what happened; how injuries are received. The tales these movies tell are often remarkable. The blows sustained in so brief a time would amaze you."[91]

Before resuming our discussion of the development of scientific crash-testing, we might take a moment to consider Dye's *Woman's Day* article in relation to J. C. Furnas's "—And Sudden Death," published in *Reader's Digest* in 1935. This brief detour will enable us to further elucidate the conceptual and discursive differences between prewar and postwar regimes of automobile accident.

Furnas's article reflected and vividly expressed the dominant discourse of accidentality in the early days of automobility. Conjuring a phantasmagoria of car wrecks, crash wounds, and fresh corpses, it explicitly indicted the vehicle operator while implicitly exonerating the vehicle engineer, the vehicle manufacturer, and the vehicle itself—to say nothing of the wider sociotechnological system in which each of these actors/agencies was embedded. It addressed its sickening depictions and strident admonitions to the reader-as-driver in an attempt to discourage him or her from behaving recklessly on the roadways. It implied the possibility of accelerated mobility without mistakes, of speed without victims, and it made the case for mindful motorism not in some "niche" forum, not in some special-interest publication on the margins of American society and culture, but in *Reader's Digest*, a mainstream, mass-circulated periodical for the general reader. To be sure, Furnas's target audience was not the nation's small, incorrigible contingent of joy-riders and thrill-seekers, but rather the nation's much more encompassing class of everyday motorists.

Dye, too, targeted a mass, middle-class, socially traditional audience, albeit a more gender-specific one. His article "Cornell University Tests Show Just What Happens in a Crash . . . and How to Protect Yourself" was the cover story in the

November 1954 issue of *Woman's Day*, a magazine geared mainly to the cultural priorities and proclivities of white, suburban housewives. The article consisted of two interrelated parts. The main part, denoted in the pre-ellipsis portion of the piece's title, offered a nontechnical overview of Cornell Aeronautical Laboratory's automotive crash-safety research program—its origins, motivations, procedures, technical apparatus, and scientific findings. The subsidiary part, denoted in the post-ellipsis portion of the title, showcased the objectified, commodified fruits of that research program—namely, Cornell-designed frame-anchored front and back seatbelts, a "chest-guard" cushion for the steering wheel, and Ensolite foam padding for the dashboard.[92] It also served as a step-by-step how-to manual, since the seatbelt kit, steering-wheel cushion, and dashboard-pad materials all had to be installed by the consumer herself—or, more probably, by her husband, a likelihood confirmed by the accompanying photographs, which depicted a man doing the work of installation.

Like Furnas, Dye wrote in the second person. But whereas Furnas had urgently warned the reader to resist the temptations of reckless driving, Dye calmly reassured the reader that a reputable center of scientific research was working hard to protect her and her family, including her young children, from the injurious effects of traffic accidents. "At our laboratory," he said, "we are finding out a great deal about automobile accidents and what happens to the men, women, and children involved."[93] Given the composition of the magazine's readership, Dye's reference to children was, no doubt, strategic. The cover photograph played on the idea that children were "precious cargo" that could be damaged or destroyed like goods in transit—or, conversely, that could be "packaged" for protection. The photo showed a woman, one hand on the wheel, the other on the gear shift, in the driver's seat of a car. Sitting next to her was a smiling young boy, unbelted and facing backward. Diagonally across his torso, written in a stenciled font like that used to label containers of goods packaged for shipment, appeared the word "FRAGILE." In the text of the article, this emotional appeal to mothers with young children found reinforcement in disturbing descriptions of crash tests involving a child-sized dummy: "Half Pint took a nose dive into the back of the front seat, sailed over that to strike his head on the steering wheel, bounced from that into the windshield and finally dropped to the floor, severely bending his back—in the wrong direction."[94]

Descriptions of this sort were only occasionally employed. In contrast to Furnas, who had pitilessly, almost fiendishly, assaulted the *Reader's Digest* reader with a barrage of grisly imagery, Dye chose to soothe the *Woman's Day* reader with a balm of scientific certainty. "You can't guard against danger until you know precisely what the danger is," he declared.[95] Fortunately, a highly skilled team of engineers and researchers is working behind the scenes to know precisely what that danger is, and how to guard against it—this, in a nutshell, was the article's thesis. Indeed, Dye promised the reader that experimental science was pinpointing the previously under-examined causes of crash injuries, and that the consumer marketplace, in coordination with the research university, was manufacturing new auto-safety products to "passively" alleviate them. "As more and more knowledge is assembled, we are becoming more and more encouraged," he enthused. "It is possible, our findings indicate, to make very substantial improvements in safety without radically changing automobile design."[96] Rather than emphasizing the need to drive safely and slow down, Dye emphasized the need to buy wisely and buckle up. Rather than proposing that accidents be eliminated through the mass correction of bad behavior, he proposed that they be "tamed" through a combination of science, engineering, and the mass consumption of passive restraints: "The chances of getting involved in an accident increase all the time: more cars are crowding our highways. All these cars are guided and controlled by humans like us, who can make errors in judgment. So there are bound to be accidents. Thus, a successful attack on this very big problem has to focus on the accident itself, the crash, the smashup. If that can somehow be tamed, injuries can be ameliorated and loss of life reduced."[97]

Dye was right to see Cornell Aeronautical Laboratory's automotive crash-safety research program as an attempt to "tame" the accident. What he did not see, however, or at least did not mention—but which his article's rhetoric and tone suggested at every turn—was that other, equally important aspect of the postwar project to tame the accident: the public-pedagogical aspect. Like the heroizing mass-media accounts of John Paul Stapp's rocket-sled rides noted earlier, and like the new breed of driver's education crash-safety films that emerged around the same time, Dye's article sought to acquaint citizens and consumers with the science and technology of crash-safety research, and to do so in a way that was at once dramatic and digestible, imperative and agreeable.[98] Stapp's media celebrity, crash-safety films' pedagogy, Dye's historiography and how-to instructions—each was part of a broader cultural and institutional impulse to domesticate high-speed accidents, to bring them "closer home," as Furnas had said of his very different approach to the problem two decades earlier. Each played a role in the normalization and popular dissemination of the notion that, while "we" (society) cannot get rid of such accidents, "we" (scientists and engineers) can manage them and even make them useful, render them socially and technologically productive. "We" can *learn from*

Cornell University tests show just what happens in a crash...

To THE DRIVER
Chest hits steering post...head smacks high on windshield

To RIGHT-FRONT PASSENGER
Body hurtles forward...head hits windshield, then dashboard

DRAWINGS BY TOM KNITCH

To A CHILD IN BACK SEAT
Nose-dives into top of front seat...head strikes front-seat passenger or windshield and dashboard

First page of Edward Dye's article in *Woman's Day* (November 1954). Note the visual allusion to frame-by-frame high-speed film analysis.

Author (right) shows William Whitlock of Woman's Day Workshop how padding can be applied over dashboard for greater protection

and how to protect yourself

For the 3 safety devices developed by Cornell University, see the following pages

By EDWARD R. DYE
Head, Industrial Division, Cornell Aeronautical Laboratory, Inc.

CARS are better mechanically today than ever before: they steer better, the brakes are better, they have more power. Yet the accident situation remains tragically bad. Every day, we read about people killed or horribly injured. The figures for a year are appalling: about 35,000 of our citizens are killed, and about 1,500,000 are injured.

In part, this is because the chances of getting involved in an accident increase all the time: more cars are crowding our highways. All these cars are guided and controlled by humans like us, who can make errors in judgment. So there are bound to be accidents. Thus, a successful attack on this very big problem has to focus on the accident itself, the crash, the smashup. If that can somehow be tamed, injuries can be ameliorated and loss of life reduced.

Thick Man and Half Pint

At our laboratory we are finding out a great deal about automobile accidents and what happens to the men, women, and children involved. We get our data by reproducing accidents in such a manner that the action of the passengers, during the violent stop, can be carefully studied. Our accident victims are dummies, especially two veterans called the Thick Man and Half Pint. The Thick Man accurately represents an adult; little Half Pint is the size of a six-year-old child. They are so constructed as to behave, in these violent adventures, exactly as living passengers would.

Except, of course, that they can't scream. There isn't much time for screaming, anyway. Everything takes place very fast in an auto accident. If a car going 20 miles an hour strikes something solid, the impact period [Continued on page 85]

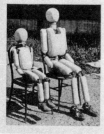

Cornell's accident victims, Half Pint and the Thick Man (left), are put through crash tests like these (right). Car, with left door removed for visibility, is steered by remote control, then stopped violently to simulate a smashup. Special cameras record what happens to movie passengers Half Pint and Thick Man

Second page of Edward Dye's article in *Woman's Day* (November 1954).

Cover of *Woman's Day* (November 1954).

accidents so that "you" can *live with* them: the promise and mythology of progress in the age of forensic reason and imagination.

A number of organizations besides the U.S. Air Force's Aeromedical Field Laboratory at Holloman and Cornell University's Aeronautical Laboratory in Buffalo were conducting scientifically controlled car-crash and force-impact tests during the 1950s. These included Wayne State University in Detroit, the Indiana State Police Department, the U.S. Army at the Aberdeen Proving Ground in Maryland, and automobile manufacturers such as Ford Motor Company. Amplifying this new wave of crash-test research was James J. Ryan's program of experiments at the University of Minnesota which, for a few years during the 1950s, operated under contract with Stapp's Automotive Crash Forces project. Ryan—the selfsame mechanical engineer and professor who invented the General Mills Ryan Flight Recorder—was contracted in December 1955 to oversee the research and development of two auto-safety devices of his own design: an energy-absorbing hydraulic bumper and a rollover bar for open-top military vehicles. In 1957, at the Third Annual Automotive Crash Research and Field Demonstration, Ryan tested his devices in front of an audience of experts. First, an Air Force weapons carrier containing an anthropomorphic dummy was crashed into a fixed barrier at twenty-five miles per hour; neither the vehicle nor the occupant was damaged. Next, Ryan and one of his graduate students drove a specially equipped passenger car into a fixed barrier at twenty miles an hour (let that sink in for a moment). Despite the fact that the car's collapsible steering wheel failed to function properly, Ryan received only a minor cut.[99]

Ryan was doubly implicated in the postwar effort to learn from mechanized-transportation failures. If Stapp was the bridge that spanned from aeronautical to automotive crash-safety research in the 1950s, then Ryan was the colossus who stood astride the two. Ryan did not really create the first flight-data recorder, although he is sometimes credited with having done so. Yet his "indestructible" instrument did constitute the most sophisticated and successful response to the Civil Aeronautics Board's call for an onboard, crash-protected graphic-recording mechanism during the postwar period, establishing technical standards (in fact, a whole program of technical and cultural development) for flight-data recorders for years to come. Ryan's work on automotive safety tackled the problem of the accident from a different vantage. Whereas his flight recorder worked on the accident after the fact as a means of preventing its future recurrence, his hydraulic bumper and rollover bar were designed to offer protection during the tiny temporality of the accident's irruption. Of these three technologies, then, only the General Mills Ryan Flight Recorder was a forensic medium in the sense defined in this book, the others being examples of what I have referred to as "single-tense safety devices." Still, Ryan is of interest here because he, like Stapp, personified—and his inquiries and innovations crystallized—the mid-twentieth-century desire to subject aircraft *and* automobile mishaps to rigorous scientific experimentation and analysis, so as to turn the accident of accelerated mobility into an object of forensic knowledge and institutional control.

Capturing Collisions on Film

If De Haven's research on impact biomechanics prepared the ground for Stapp's rocket-sled experiments, those experiments, in turn, paved the way for the work of scientists and technicians at UCLA's Institute of Transportation and Traffic Engineering (ITTE). Established in 1947 by an act of the California State Legislature, the ITTE was commissioned to study the design, construction, operation, and maintenance of U.S. airports, highways, and related public transportation facilities. By 1960, the ITTE had conducted nearly fifty automobile collision experiments, the purpose of which, according to the chief research engineer Derwyn M. Severy, was to provide "critically needed data on physical factors relating to vehicular collision dynamics and attending motorist injuries."[100] High-speed cinematographic techniques and technologies, along with an array of sensing devices, electrical transducers, and oscillographic recorders, enabled Severy and his colleagues to mediatize and scientifically scrutinize the instant of impact, that moment of maximal danger (to bodies) and destruction (to machines).

The ITTE first forayed into crash-testing in 1950, when all that was known about the dynamical characteristics of automobile collisions was based on the "educated guess" of engineers "rather than on experimental findings."[101] Having learned that a troupe of professional stunt drivers performed head-on collisions as part of its show, ITTE researchers obtained permission to film the spectacular smashups in the interest of scientific inquiry. Two studies were made in Southern California with the cooperation of the Joie Chitwood Auto Daredevils, the first in June 1950 at the Culver City Speedway, the second in June 1951 at the Carrell Speedway in Gardena.[102] In each case, a single Eastman high-speed film camera was trained on the car approaching the predicted point of impact from screen right. Timing devices, measuring boards, and other reference markers were placed in view of the camera. These manipulations of mise-en-scène, in conjunction with a Bausch and Lomb contour-measuring projector, facilitated the frame-by-frame analysis of the automobiles' movements (their material displacements and deformations) throughout the collision process. The drivers' movements (their bodily trajectories and, possibly, traumas) were not amenable to such analysis, however, as the stunt required the daredevils to duck for cover—and, in effect, disappear—behind the front-seat assembly in the seconds immediately prior to impact. Partly for this reason, the tests were of limited scientific value. Contours that could not be registered by the camera could not be measured with the projector, and contours that could not be measured with the projector were, as far as the ITTE was concerned, no contours at all. Crouched on the backseat floor, removed from the field of technological vision, the motorists' bodies remained as enigmatic and inaccessible as ever. From a research standpoint, the cars might as well have been occupied by ghosts.

Compounding the problem, in Severy's estimation, were "the difficulties encountered in maintaining adequate experimental control over [the] crashes."[103] The speeds of the vehicles, the angles of approach, the points of mutual contact, the site and time of impact: such unregulated variables served to unsettle an idealized order, injecting an unwelcome element of chance into the rational-scientific scheme of things. That which enhanced the excitement of thrill-seeking spectators—namely, the collision's material and spatiotemporal contingencies, its *accidentalities*—undermined the aims of truth-seeking re-

Head-on collision experiment using stunt drivers. Courtesy of UCLA's School of Engineering and the Regents of the University of California.

searchers. The popular taste for spectacle hampered the pursuit of scientific knowledge; the allure of indeterminacy worked against the need for certainty. "While we learned a great deal and received splendid cooperation from both management and drivers," Severy remarked, "we were eventually forced to the conclusion that the demands of the audience and our research requirements were incompatible."[104]

Following the stunt-driver tests, the ITTE retreated for a time from the realm of experiment and, in a move that recalled De Haven's wartime research, turned its attention to the investigation of actual roadway accidents. Working in tandem with the Los Angeles Police Department, Severy and his colleagues inspected and photographed crash scenes, took statements from motorists and eyewitnesses, and developed case files of "typical accident types."[105] These methods, illuminating in some respects, proved to have significant shortcomings. "The problem of obtaining factual information from post crash observations is an especially deceptive one," Severy lamented. "Not only is it generally impossible to determine with reasonable accuracy the pre-impact velocity, but it is also generally impossible to be certain that post accident observations are truly representative of what actually happened."[106] Trained accident investigators, arriving on the scene after the fact, were not the only ones afflicted with the inability to truly represent "what actually happened," according to Severy. When they were not too "shocked" to speak coherently, motorists involved in collisions frequently "slanted [their statements] for self protection," and eyewitness testimonies, despite being comparatively dispassionate, were often confused or mutually contradictory. Laypersons were found to be fallible psychologically, morally, and perceptually; they did not always think straight, talk straight, or see straight. The film camera, by contrast, was held to have no moral frailties or psychoperceptual failings, and it soon became clear to ITTE engineers that, if they were to comprehend the instant of impact in all its complexity, they would again have to enlist the superhuman services of high-speed cinematography.

The ITTE returned to full-scale collision experimentation in February 1954.[107] Previous attempts to penetrate the mysteries of the car crash had been only partially successful. The speedway studies had ceded too much control to crowd-pleasing daredevils who obscured themselves right when their photographic visibility mattered most, whereas the accident investigations had relied too heavily on the post hoc guesswork of experts and on the imperfect perceptions and recollections of ordinary individuals. This time around, all aspects of the experiments—visibilities, velocities, geometries, temporalities—would be meticulously planned, tightly managed, and carefully monitored. Opportunities for electronic and photographic data generation would be maximized; the potential for human error or interference would be minimized. The quest for truth and certainty would give no ground to accidentalities or indeterminacies. Nothing would be left to chance.

In two separate experiments, a 1937 Plymouth sedan was equipped with accelerometers and seat-belt tensiometers connected by cable to a twenty-four-channel recording oscillograph mounted on a mobile instrument truck. A pair of specially developed anthropometric dummies, with weight distributions "comparable to [those] of man" and joint fixations "closely resembling [those] of a person forewarned of an impending collision," occupied the front seat, simulating driver and passenger.[108] Wired with electrical sensors, the dummies were subjected to four different conditions of restraint (two per test): one wore a lap belt, another a chest belt, another a shoulder belt, and one wore no belt at all. The sedan was pushed from behind by the instrument truck for several hundred feet and then allowed to coast, at twenty-five miles per hour, into a wall of utility poles backed with earthwork. An assortment of cameras, variously positioned, captured the action for subsequent review and evaluation. "A good

Head-on collision experiment using anthropometric dummies. Courtesy of UCLA's School of Engineering and the Regents of the University of California.

deal of the data for the tests are contained in these films," an ITTE technician wrote in a piece for *California Engineer* in 1954.

Not only do the pictures (taken at speeds between 1000 and 2400 frames per second) reveal deceleration rates of the car and dummies, but also deceleration patterns of the car frame by means of small metal targets mounted on rods which are in turn welded at several points along the frame.

Two GSAP [gun-sight aiming point] movie cameras mounted on the shelf behind the back seat of the crash-car revealed additional information on the dummies' motion during impact. Another camera located thirty feet from the barrier panned the movement of the car and truck as they sped toward the barrier. A similar motion picture camera mounted on top of the instrument truck, trained forward on the car and barrier, points up steering problems and crash behavior.

A speed graphic camera with an electronic timing device took a still picture of the car two-hundred milliseconds after the car first contacted the barrier. This was the time of zero forward velocity, approximate maximum deceleration, and greatest crush-in volume.¹⁰⁹

The cumulative photographic register contained what another ITTE technician called "a complete time history of the motion during impact and deformation period," a full, faithful chronicle, in pictures, of the instantaneous activity of automobile catastrophe.¹¹⁰

These unorthodox experiments attracted the attention of at least two popular publications in 1954, both of which accentuated the tests' destructive spectacle as well as the dummies' uncanny surrogacy and violent sacrifice. "A simulated man lost his head yesterday in a prearranged automobile accident," reported the *Los Angeles Times*. "Another 'man' who was 'driving' the car will be studied by experts to determine what injuries he might have received if he had been real."¹¹¹ Under the ominous headline "Dummy 'Killed' in Smashed-up Car Could Be You," *Popular Science* described the situation with mock solemnity: "A car going 25 miles an hour smashed into a solid barrier of telephone poles the other day. The front-seat passenger was killed. His ribs were broken, his

Two hundred milliseconds after impact. Courtesy of UCLA's School of Engineering and the Regents of the University of California.

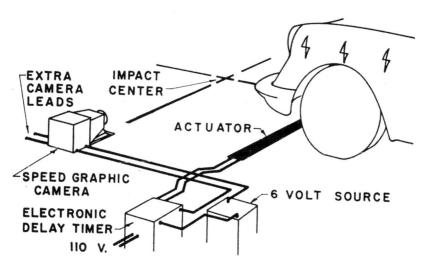

Equipment for precision-timed photographs. Courtesy of UCLA's School of Engineering and the Regents of the University of California.

neck fractured, his vertebrae driven together. The driver escaped with cracked knees." And then, the punch line: "Fortunately, the occupants of this car, deliberately cracked up, were dummies."[112] Each article was accompanied by a vivid photograph of the sedan smashing into the barrier, its front end crushed and crumpled, its rear wheels raised off the ground like the hind legs of a bucking bronco (an image completed by the instrument cable, which swung from the back of the car like a whiptail). A post-crash photo of the dummies in *Popular Science* bore the caption: "One 'dies,' one is saved by shoulder harness." The same magazine noted that numerous electronic and photographic instruments had been pressed into service to tell "the story of what happened to the dummies and to the car."[113]

By turns expositive, macabre, and ideologically reassuring, these and other press accounts (including the ones discussed in the previous section), together with a series of driver's education films produced by the ITTE beginning in 1957, introduced the postwar American public, verbally and visually, to the brave new world of scientific crash-testing. It was a world in which vehicles collided at high speeds and caused damage neither playfully nor haphazardly but ceremoniously and precisely; in which bodies with human form were forcibly tossed and concussed yet experienced no fear and endured no pain; in which the legalistic search for the cause of the crash and for someone to morally blame was superseded by the medicalistic search for the causes of crash injury and for something to technologically improve.

That high-speed cinematography was indispensable to the success of this latter search was recognized as early as 1946, when aviation-medicine researchers at the Naval Medical Research Institute in Bethesda, Maryland, used an Eastman camera with an exposure rate of 3,000 frames per second to reveal the "physical and physiological reactions of human subjects to forces on the impact decelerator."[114] (A curious contraption, the impact decelerator used falling weights and a sixty-five-inch steel rod to sharply tug a safety harness against the torso of a test subject as a means of mimicking the impact forces acting on an airplane pilot during a crash landing.) As indicated in an Air Force technical report published in 1951, Stapp's rocket-sled experiments at Edwards, too, had sought to mobilize the revelatory power of the moving image:

> The subject's movements during impact were recorded in frontal view by a 16 mm. 128 frame per second movie camera mounted on a tripod and focused through a tubular housing passing through the windshield. Aft facing runs were similarly covered by mounting the tripod at the rear of the sled. Profile studies of displacement during passage through the braking area were obtained by a battery of six Eastman High Speed movie cameras operated at approximately 1400 frames per second. The cameras were spaced to give overlapping coverage of ten to twelve feet sectors of the braking area.
>
> The entire braking area was covered in one field by a ribbon frame camera operating at 120 exposures of 1/10,000 second duration per second.[115]

To enable the photographic measurement of displacements of various kinds, numbered reference markers, alternating in black and white, appeared on the sled architecture, on a metal baseboard that ran the length of the braking area, and, in some experiments, on the face, neck, shoulder, and knee of the test subject himself.

These pioneering applications of high-speed cinematography to crash-injury research provided not only a methodological template but also an epistemological rationale for the ITTE's increasingly sophisticated use of photographic apparatus throughout the 1950s. Like their predecessors in aviation medicine, ITTE researchers put their faith in the film camera, trusting in its indexically guaranteed ability to "see" the unseen and otherwise unseeable properties and processes of extreme force impact. "Many collision events occur too quickly for even a group of trained observers to perceive them," Severy said in a speech

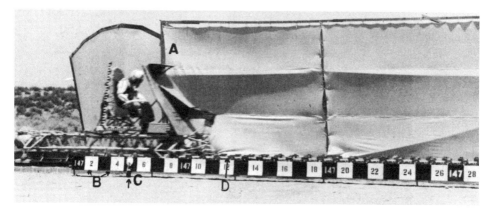

Rocket-sled experiment showing (a) photographic backdrop, (b) foot markers, (c) stroboscopic timer, and (d) brakes. Source: John Paul Stapp, "Human Exposures to Linear Deceleration: Part 2. The Forward-facing Position and the Development of a Crash Harness," *Air Force Technical Report 5915*, December 1951.

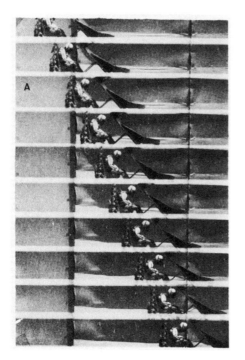
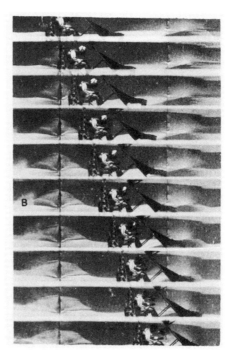

Ribbon-frame camera captures rider displacement during rocket-sled deceleration. Source: John Paul Stapp, "Human Exposures to Linear Deceleration: Part 2. The Forward-facing Position and the Development of a Crash Harness," *Air Force Technical Report 5915*, December 1951.

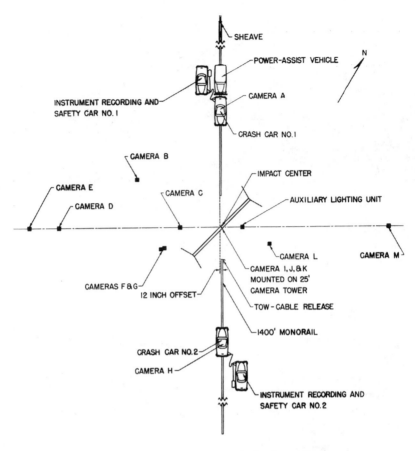

(*above and opposite*) Photographic instrumentation for head-on collision experiments. Courtesy of UCLA's School of Engineering and the Regents of the University of California.

delivered at the California State Governor's Safety Conference in 1954. "Only high speed cameras can provide such information on a time basis expanded sufficiently to allow the human mind to perceive the events."[116]

Severy elaborated on the time-expanding, perception-enhancing capacities of those cameras in a paper he presented at the Hollywood Section meeting of the Society of Motion Picture and Television Engineers in November 1957. Published three months later in *Journal of the SMPTE* under the title "Photographic Instrumentation for Collision Injury Research," the paper explained in exquisite detail, over nine copiously illustrated pages, how the ITTE staged, shot, and, with the aid of an optical comparator, scrutinized its crash tests.

It also explained the manifold difficulties involved in doing so. Because the

Camera	Position (see Fig. 3)	Type of film	f setting	Frames/sec, or shutter speed	Distance, (feet)
High-Speed (ITTE) Eastman II 63mm.........	D	Superior 4* (Du Pont)	2.0	1400	120
High-Speed (ENGR) Eastman I 63mm........	E	Superior 4	4.0	900	138
High-Speed Fastax I 50-mm..................	I	Superior 4	2.7	1400	24
High-Speed (AEC) Eastman II 63mm.........	F	Superior 4	4.0	600	38
High-Speed Fastax II 25-mm..................	M	Superior 4	2.8	1400	123
Mod. High-Speed (GSAP) 12.5mm...............	H-(Nash)	Tri-X Neg. (Eastman)	4.0	200	5.5
Mod. High-Speed (GSAP) 25mm................	A-(FORD)	Tri-X Neg.	4.0	200	6
Tower GSAP 17mm.....	J	Plus X (Eastman)	8.5	64	24
Tower Time-Delay† 127-mm.................	K	Royal Pan (Eastman)	11.0	1/400	24
Ground Time-Delay† 127-mm.................	G	Royal Pan	11.0	1/400	10
Ground Time-Delay† 127-mm.................	C	Royal Pan	11.0	1/400	25
Cine-Pan 25mm.........	B	914A Du Pont	16	24	60
Cine-Pan Anamorphic Lens................	L	914A Du Pont	16; #1 filter	24	31

*Superior 4, Du Pont 928A Negative (footage numbered).
†Speed Graphic cameras controlled by electronic time-delay device.

car crash was, in Severy's pithy locution, "an extremely complicated phenomenon of very brief duration ending in destruction"—because, in other words, so many and sundry things happened so fast and with such violent finality—an enormous amount of time, effort, and resources was needed to re-create it exactly and to record it accurately.[117] "Experiments which eat up research funds at a rate of $40,000 a second require extremely careful planning and particular attention to operational details to minimize the chance of loss of data through any one of a multitude of system failure potentialities."[118] Scientifically controlled smashups are huge (and hugely expensive) undertakings, begun and ended in the blink of an eye, and thus every step has to be taken, every safeguard put in place, to reduce the risk of something—anything—going wrong. Mistakes and malfunctions haunt even the most coolly instrumental, the most rationally efficient of spaces: the specter of accidentality always and everywhere looms. This, the same anxious awareness that called the collision experiments into being in the first place—an awareness codified in the cultural imagination as Murphy's Law—was recapitulated in the experimental procedures themselves.[119] "If during the run a malfunction occurs," Severy wrote, "the instrumentation engineers may stop the collision vehicles by air brakes operated remotely from their instrument cars."[120] A similar precautionary impulse underlay the ITTE's film-processing practices: "Following an experiment the many rolls of expended film are divided into two carefully selected groups so that each group maximizes the photographic coverage. These groups are processed by the laboratory independently so that an unfortunate error would not result in the destruction of irreplaceable photographic data."[121]

In the 1950s, as now, professional engineers believed that the surest way to defend against the possibility of ruinous accident was to build "redundancy" into the system. By design, a specified task was to be done once and done again, and perhaps again *again*, each time somewhat differently. Fail-safety is realizable through preemptive reiteration, disaster avertable by "backing up the data" beforehand: at the ITTE, this rationalist strain of wishful thinking was evident in the conceived relationship between electronic and photographic apparatus. In order "to reduce the possibilities of loss of primary data from failure by a single instrumentation recording system during collision," Severy explained, "both photographic and electrical-transducer instrumentation are applied as extensively as economically feasible and with sufficient overlapping of the two systems."[122] ITTE engineers were particularly worried that certain electrical transducers were susceptible to failure: "Transducers to be mounted near collapsing structures are encased in a rigid steel box. Even with this precaution the box may be bent, the electrical lead severed, punctured, or the entire box torn loose if it is situated too close to the grossly collapsing structure."[123] Operating at a safe remove from sites of structural collapse, from the crash's implosive, disintegrative ground zero, film cameras were expected to function reliably even, and especially, if their electronic counterparts failed completely. Yet, no matter how reliably those cameras functioned, the "continuity of a photographic sequence [could still] be interrupted by flying debris or the incorrectly anticipated movements of the subject."[124] Redundancy be damned: the interruptions of noisy contingency were impossible to repress.

The issue of failure notwithstanding, each system of instrumentation was thought to have its advantages and privileged sphere of application. Electrical media were good at sensing and measuring strains and stresses, tensions and pressures. Photochemical media were good at spying and mapping the changing positions of anthropometric bodies in motion as well as the changing formations of automobile bodies under destruction. In a two-car collision experiment, for example, a mere ten feet of film from a single overcranked camera was said to provide a wealth of data on the motorists' postural variations, the vehicles' plastic and elastic deformations, and the motorists' and the vehicles' acceleration and deceleration patterns.[125] Celebrating these and other cinematographic feats in his *Journal of the SMPTE* article, Severy proclaimed that "photographic close-ups of the rapid sequences of injury-producing events have provided new and otherwise unobtainable scientific insight into the mechanisms of injuries in these events."[126]

The Destructive Split Second

The Institute of Transportation and Traffic Engineering was not the first organization in the United States to point a film camera at a staged car crash for reasons other than popular entertainment. In the early 1930s, General Motors Corporation began using its massive proving ground in Milford, Michigan, to conduct three kinds of destructive experiment: level rollover, spiral-ramp rollover, and barrier impact.[127] In a typical level rollover test, a car was overturned by driving it into a skid on a flat sod field. In a typical spiral-ramp rollover test, a driverless car was pushed onto a helical ramp located at the top of a hill, causing the vehicle to turn over and tumble down. And, in a typical barrier-impact test, a driver, standing on the left-side running board, steered a car toward a concrete wall and jumped off just before impact. Extant footage from the General Motors archives suggests that a lone film camera, running at standard speed and framing a long shot, was, if not routinely, then at least occasionally on hand to record the experiments.[128]

According to *General Motors Engineering Journal*, the objective of the company's early crash tests was "to evaluate the integrity of the body structure," on the one hand, and to supply "information on collision impacts under controlled conditions that could be correlated with highway accident damage," on the other.[129] Significantly, the experiments were designed to assess the structural integrity of the automobile body, not the vulnerability of the human body. At stake was the car's durability, not the occupant's safety or survivability (the latter term, tellingly, did not gain currency in transport-safety circles until the 1950s). Rather than concerning the damage done to living beings, the task of correlating test results with highway accidents concerned the damage done to lifeless machines. Dummies, consequently, were deemed unnecessary. Drivers, when required, fulfilled a purely instrumental function: speed and steer the vehicle. What happened to motorists upon impact was of no experimental import. Denied even a presence by proxy, the human body was written out of the crash scenario, expelled from the scene of extreme force impact. As for the film camera, its duty was simply to document the event, to establish its actuality for the archival record, not to facilitate its optical analysis. General Motors engineers of the 1930s did not pore over crash-test footage, analyzing it frame by frame with scientific intention and mathematical precision; instead, they watched a given experiment unfold, and inspected the wreckage afterward, with nothing but an "unaided eye."[130]

No timing devices, no measuring boards, no reference markers. No electrical sensors or transducers, no oscillographic recorders, no high-speed cameras, no contour-measuring projectors. No painstaking exactitudes, no systemic redundancies. And no experimentally important human or humanoid bodies. This is what cutting-edge crash-testing looked like in the days when the discourse of the reckless driver reigned supreme.

The ITTE collision experiments of the 1950s represented the supremacy of a very different discursive regime—which is to say, a very different approach to the problem of accidents. "This approach stemmed from the concept that a certain number of automobile collisions are inevitable but that the resulting injuries are, to a large extent, preventable," Severy declared. "This concept is not in conflict with the statement that injuries are prevented by preventing accidents. It simply recognizes the limitations of accident prevention measures and deals directly with the sequential important issue, the reduction of motorist injuries."[131]

These comments, made in the late 1950s, encapsulate the conceptual and discursive underpinnings of the postwar turn to crash-injury medicine and its technological corollary, crashworthiness engineering. Despite more than a half-century's worth of widespread accident-prevention efforts based on disciplining inexperienced drivers and punishing incautious ones, car crashes continued to plague America's streets and highways, with no end in sight. Accidents, once thought eliminable through the twin solution of education and law enforcement, now seemed "inevitable," an unfortunate fixture of modern life. This did not mean, however, that nothing could be done to lessen their menace. If the curse of roadway contingency could not be lifted completely, it was still possible, supposedly, to rob it of much of its deadly power. For Severy and his like-minded contemporaries across the country in industry, academia, and the military—most notably, Alex L. Haynes at Ford Motor Company in Detroit, Edward R. Dye at Cornell Aeronautical Laboratory in Buffalo, and, in yet another phase of his remarkable career, John Paul Stapp at Holloman Air Force Base's Aeromedical Field Laboratory in New Mexico, all of whom directed automotive crash-research programs during the 1950s—the pressing issue was not accidents per se, but rather their harmful effects on human anatomy. The only way to mitigate those effects, according to Severy and the aforementioned others, was to understand the origins and mechanics of crash injury. And the only way to truly understand those origins and mechanics was to conduct scientific experiments using state-of-the-art electronic and, above all, photographic apparatus.

The ITTE's use of photographic apparatus to analyze fast-moving phenomena participated in a tradition of technically mediated scientific looking that recalled Etienne-Jules Marey's and Eadweard Muybridge's animal-locomotion studies as well as Frank and Lillian Gilbreth's time-motion studies.[132] As Severy himself noted, "Close-up observation of split-second events has long been a recognized application for high-speed photography."[133] There was, nevertheless, something novel and unusual afoot at the ITTE (and at Ford, Cornell, and Holloman) during the 1950s—something peculiar about the phenomena being analyzed, something distinct about the "split-second events" being observed. Through the agency of high-speed cinematography, a previously hidden, culturally repressed aspect of accelerated mobility was suddenly on view, and it proved to be as psychologically disturbing as it was visually striking and forensically enlightening. It came to be called "the second collision."[134]

Prewar auto-safety discourse acknowledged only the most obvious collision in an accident: that between the vehicle and an object outside it. The car crash was conceived simply as a chance meeting of two entities external to each other, and the moment of contact between those entities—the instant of impact—was considered undeserving of attention and, in any case, forbidding in its opacity. The destructive split second was held to be undivided, internally undifferentiated, and so dense as to be without duration. Postwar auto-safety discourse, by contrast, identified and emphasized an additional, less obvious collision: that between the vehicle and an occupant inside it. This chronologically second collision, occurring mere milliseconds after the first, was the collision of urgent import, as it was the one that maimed or killed. The initial, exterior impact was bad for the machine (it produced wreckage), but it was the slightly subsequent, interior impact—body against steering wheel or dashboard or windshield—that was bad for the motorist (it produced carnage). With the discovery and discursive privileging of the second collision, the crash verily became the "extremely complicated phenomenon" of Severy's description. No longer a unitary, uniform, effectively atemporal thing, it was now a dynamic, durational multitude of causally connected things, a heterogeneous assemblage of masses and forces developing, interacting, and transforming over a delimited period of time. Formerly dismissed as either uninteresting or unapproachable,

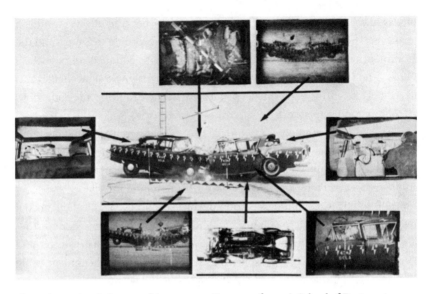

Three-dimensional photographic coverage. Courtesy of UCLA's School of Engineering and the Regents of the University of California.

the instant of impact was now, for the first time, embraced as a source of fascination susceptible to forensic mediation and rational explanation.

Examined in slow motion, that instant was divisible into two sequential segments: the interval between the first collision and the second, and the interval between the second collision and the terminus of the collision process, the final cessation of all motion. The car crash was considered a properly human tragedy because of the brutally humbling events that took place during these tiny temporal intervals. As ITTE researchers were keen to point out, accidents frequently exposed motorists to tremendous deceleration forces—forces that rendered them incapable of resistance—in an extremely compressed span of time. "No greater duration than one-quarter of a second is required for two well-engineered automobiles approaching one another at 50 mph to be converted into junk," Severy observed. "Driver reaction time—the time required to perceive, interpret and commence to react to an environmental disturbance—averages 3/4 sec. Should a head-on collision between cars traveling 50 mph appear imminent when the cars are still 100 ft apart, no preventive effort could be initiated before they crashed. The ensuing quarter-second of destruction embroils the motorists within collapsing structures with injurious forces having a magnitude measured in tons rather than pounds. Even if there were sufficient time to brace himself, these forces completely overwhelm the motorist."[135]

Precision-timed photograph taken from camera tower. Courtesy of UCLA's School of Engineering and the Regents of the University of California.

The villain of prewar auto-safety discourse, the reckless driver, is conspicuously absent from this grim account of automobile catastrophe. Absent, as well, is the moral scheme that endowed his or her actions and volitions with meaning and consequence, the ethico-juridical logic that assigned him or her responsibility and insisted on his or her accountability. Just as the question of culpability goes unasked here, so the prospect of behavioral reformation—and of accident prevention—goes unimagined. For the collision experimenter, such considerations are plainly beside the point; the ethics of automobility necessarily take a backseat to the physics of accidentality. Surrounded by "collapsing structures," overpowered by "injurious forces" of monstrous proportion, the motorists in Severy's disaster scenario are denied the opportunity to determine their own fates, deprived of the ability to make a difference. They cannot possibly anticipate or assess, much less respond to, the onrushing developments in so short a time. They can do nothing to deliver themselves from their perilous predicament, to alter the inexorable course of events in which they are haplessly, hopelessly, "embroiled." Neither are their perceptions acute enough, nor their reactions quick enough, nor their bones, sinews, and muscles strong

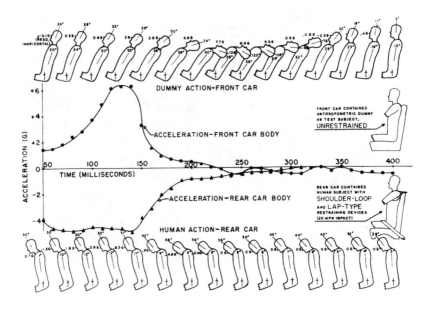

Kinematic analyses derived from high-speed motion pictures. Courtesy of UCLA's School of Engineering and the Regents of the University of California.

enough. The accident condemns them to impotence and incapacity: it all happens too soon; it is already too late. Trapped inside a rapidly decelerating, violently imploding machine, the bodies of driver and passenger are reduced to the status of inanimate objects, no more active or effective than the test dummies that simulate them.

ITTE engineers, too, were restricted by bodily impotencies and incapacities brought on by the destructive split second: their eyes beheld it only blurrily; their minds remembered it only impressionistically. In a bid to overcome these mortal deficiencies, they relied on the overcranked camera and its complement, the optical comparator, to render the instant of impact legible and intelligible, riding the current of a cultural desire for the scientifically knowable, institutionally controllable accident. The strategic photographic seizing, slicing up, and slowing down of the collision process encouraged the notion that even its most complicated riddles, whether physical in nature or physiological, were capable of being deciphered. "In scientific cinema," according to Scott Curtis, "there is a double 'extraction': the camera penetrates and captures a reality otherwise invisible, and then, through quantitative analysis, useful, objective data is 'extracted' from the image itself."[136] At the ITTE, high-speed cinematography was employed to "doubly extract" the obscure reality of automobile catastrophe. Its technological vision promised to scientifically penetrate the crash's closed spatiality by capturing its most crucial temporality. It made it possible, as never before, to believe that the accident of accelerated mobility could be objectively revealed, forensically known, and, as Martin Heidegger would have it, willfully mastered.[137]

An Unscheduled Event

In January 1959, the Institute of Transportation and Traffic Engineering released a technical report titled *Auto Crash Studies*, which detailed the procedures and results of eight collision experiments.[138] Seven were planned and purposely executed: three car-to-car side impacts, two car-to-car head-on impacts, and two car-to-barrier impacts. One of the "experiments," however—a car-to-truck rear-end impact—was quite accidental:

> Following a successful series of . . . head-on collision experiments, the guide track was shifted to the side-impact configuration. Other equipment was modified along with procedural changes necessary for this new experimental configuration. Special dry runs were made to evaluate the modified system's capability for synchronization of crash vehicles before the first side-impact experiment was conducted. The usual extensive pre-crash checklist was reviewed to determine the state of readiness. Notwithstanding these preparations and precautions, the first side-impact experiment failed as a result of vehicle positions becoming asynchronous during their approach to the impact point. Loss of power by [the] towing crash vehicle allowed the towed crash car to gain headway and pass through the intersection immediately ahead of the towing crash car. When this misrun became apparent, emergency brakes were applied which stopped the power car but the towed car reached the rear of the camera truck before its brakes were energized. This highly instrumented car was wrecked.[139]

Equipment modifications and procedural changes, dry runs and system evaluations, pre-crash checklists and readiness reviews: even in redundant conjunction, all these preparations and precautions could not keep the forces of accidentality at bay. As with the legendary events that gave rise to Murphy's Law approximately a decade earlier, this scientific-experimental snafu—an acronym that, like Murphy's Law, postulates the normality of failure: *situation normal, all fucked up*—allegorizes, and ironizes, the accident's untamableness, its refusal to be banished. It mocks what Robert Castel calls modernity's "gran-

An unscheduled event. Courtesy of UCLA's School of Engineering and the Regents of the University of California.

diose technocratic rationalizing dream of absolute control of the accidental" by conjuring the nightmare of the accident's indomitability, its victorious revolt against the reign of technocracy.[140]

Eager to have the final word, the apostles of scientific rationalism at the ITTE, as though alluding to Ernst Bloch's characterization of the accident as "a production that knows no civilized schedule," dubbed the mishap "an unscheduled event," even as they transfigured it, in the pages of their report, into a comfortably coherent set of "technical findings."[141]

epilogue

RETROSPECTIVE PROPHECIES

In the generic plot of a liability trial, the world has slipped from the ordinary to the extraordinary by the short path of a passive and unfathomable slippage that is resurrected into recoverable intelligibility by being subdivided into a sequence of discrete actions. . . . Implicit in this mimesis of restorability is the belief that catastrophes are themselves (not simply narratively but actually) reconstructable, the belief that the world can exist, usually does exist, should in this instance have existed, and may in this instance be "remakable" to exist, without such slippage.
—ELAINE SCARRY, *The Body in Pain*

The important thing is that the photograph possesses an evidential force, and that its testimony bears not on the object but on time. —ROLAND BARTHES, *Camera Lucida*

Speculating in Retrospect

In 1880, Thomas Huxley delivered a lecture at Working Men's College in London titled "On the Method of Zadig." Huxley, a prominent naturalist and staunch Darwinist, claimed that Zadig, the philosopher-hero of Voltaire's novel of the same name, practiced an art of "methodised savagery."[1] Like a primitive hunter, Zadig was an astute tracker and discoverer, able to "perceive endless minute differences where untrained eyes discern[ed] nothing." From this observation, Huxley proceeded to argue that Zadig's method of tracking and discerning lay at the root not only of the science of paleontology, which deduces by "reasoning from a shell, or a tooth, or a bone, to the nature of the animal to which it belonged," but of "all those sciences which have been termed . . . palaetiological, because they are retrospectively prophetic and strive towards the reconstruction in human imagination of events which have vanished and ceased to be."[2]

Huxley's odd locution—"retrospectively prophetic"—was as deliberately oxymoronic as his invocation of palaetiology was referentially and conceptually specific. *Palaetiology* was the nineteenth-century philosopher and science historian William Whewell's neologism for "the application of existing principles of cause and effect to the explanation of past phenomena."[3] "It is perhaps a little hazardous to employ phraseology which perilously suggests a contradiction in terms—the word 'prophecy' being so constantly, in ordinary use, restricted to 'foretelling,'" said Huxley. "Strictly, however, the term prophecy applies as much to outspeaking as to foretelling; and, even in the restricted sense of 'divination,' it is obvious that the essence of the prophetic operation does not lie in its backward or forward relation to the course of time, but in the fact that it is the apprehension of that which lies out of the sphere of immediate knowledge; the seeing of that which, to the natural sense of the seer, is invisible."[4] Huxley contrasted (and thus implicitly compared) the retrospective prophet to other sorts of augur and soothsayer. Whereas "the foreteller asserts that, at some future time, a properly situated observer will witness certain events," and "the clairvoyant declares that, at this present time, certain things are to be witnessed a thousand miles away," the retrospective prophet—or "backteller"—"affirms that, so many hours or years ago, such and such things were to be seen."[5]

Suggestive in its paradox, the notion of retrospective prophecy, or backtelling, compounds a particular desire to speculate, a particular desire to see, and a particular desire to say—all predicated on an ideal and homogeneous temporality, on a linearist, continuist, reversiblist conception of time. As formulated by Huxley, retrospective prophecy articulates, in one and the same operation: the will to imaginatively reconstruct something past; the will to apprehend something absent, to grasp something not so much categorically nonexistent as contingently no-longer-existing; and the will to speak out about something immediately unknowable, something naturally invisible. Significantly, this complex operation is carried out under the aegis of modern science, a positive palaetiological science—a science authorized to speculate about past events because underwritten by the supposedly universal laws of time and causation.

In his monumental, multivolume *History of the Inductive Sciences*, first published in 1837, Whewell explains his reasons for devising the term *palaetiology*: "The sciences which treat of causes have sometimes been termed *aetiological*.... [The science of] *Palaeontology* ... treats of beings which formerly existed. Hence, combining these two notions, *Palaetiology* appears to be a term not inappropriate, to describe those speculations which thus refer to actual past events, and attempt to explain them by laws of causation."[6] The palaetiological sciences, "diverse as they are in their subjects," according to Whewell, all "endeavor to ascend to a past state of things, by the aid of the evidence of the present."[7] In other words, these sciences (among which Whewell included geology, archaeology, comparative philology, and physical astronomy) demonstrate how "the relics and ruins of . . . earlier states are preserved, mutilated and dead, in the products of later times."[8] Dead relics, mutilated ruins. In evoking such ghostly, ghastly images, Whewell, a leading exponent of scientific rationalism—in fact, the coiner of the word *scientist*—accidentally betrays the uncanny, shadowy dimensions of what he considered palaetiology's "rigorous and systematic" endeavor.[9]

Traces of uncanniness and hints of weirdness can be detected, as well, in that avowedly palaetiological repertoire of procedures Huxley dubbed "the method of Zadig." Huxley maintained that Zadig's method was premised on an eminently reasonable, empirically verifiable assumption: "that we may conclude from an effect to the pre-existence of a cause competent to produce that effect."[10] Yet, in the very same breath, he linked this ostensibly rational method to the occult arts: to prophecy, to divination, to clairvoyance. In Huxley's formulation, then, science, speculation, paranormal perception, and supernatural revelation intermingle and, indeed, become all but indistinguishable.

For Jacques Derrida, "speculation is always fascinated, bewitched by the specter." Speculation is an "alchemy . . . devoted to the apparition of the specter, to the haunting or the return of *revenants*."[11] What Derrida says about speculation in general applies with special force, or at least with singular inflection, to that peculiar mode of retrospective speculation known today as "forensic accident reconstruction."

In this epilogue, I examine some of the methods, discourses, and media technologies deployed in the contemporary science and practice of forensic accident reconstruction, specifically motor-vehicle accident reconstruction. As with all things forensic, forensic accident reconstruction is allied with palaetiology and retrospective prophecy, owing to its obsession with past causes and evidentiary traces. (I here use the word *obsession*, with its overtones of uncontrolled affect and excessive desire, to signal the prior contamination of all scientific preoccupations and pursuits.) Nevertheless, forensics has its own logics, rhetorics, and institutional articulations. Aside from its ties to the palaetiological sciences, what distinguishes the forensic impulse from other epistemophilic "instincts," other longings for understanding secured by scientific reason, is its fixation on violent crime and catastrophe, its autoptic fascination with carnage and wreckage, with biological and technological corpses. Because of this, and because of the roles played by mediation and speculation in its elaboration—

mediation being intrinsically risky, speculation being irreducibly uncertain—the forensic impulse is forever caught between the scientific rationality that underwrites it and the specters of death and contingency that haunt it.

Rendering the Scene

What is accident reconstruction? How do its practitioners define its purpose? Rudolf Limpert in *Motor Vehicle Accident Reconstruction and Cause Analysis* writes, "The fundamental objective of accident reconstruction is to determine what happened at one or several points in time."[12] Northwestern University Center for Public Safety says, "Describing the events of the collision, in appropriate detail, is the aim of collision reconstruction."[13] And another contemporary expert on the subject puts it this way: "The forensic engineer practicing in the field of traffic accident reconstruction attempts to provide insight into the essential character of such mishaps."[14] Before an accident can be forensically *reconstructed*, however—before its temporal sequence can be determined, its constitutive events described, its "essential character" discerned—the scene of its occurrence must be forensically *investigated*.

At the heart of the investigation process lie the trained recognition and skillful documentation of accident-scene data. "The objectives of accident scene data collection are the honest and accurate identification, interpretation, and documentation of scene and related data that form the basis for the reconstruction of the accident."[15] The outcome of forensic inquiry hinges on the quality and quantity of the evidence discovered at the scene. "In most cases," states the *Traffic Accident Investigators' Manual*, "the success or failure of all other segments of the investigation depends almost entirely upon the evidence gathered during the at-scene investigation."[16]

Like all forensic practitioners, the accident investigator is, first of all, a reader of traces, an interpreter of what Limpert calls "markings."[17] Tire marks, gouge marks, scrape marks, slide marks, liquid marks, human-contact marks, animal-contact marks, exterior vehicle marks, interior vehicle marks: each of these markings is, for the investigator, inherently meaningful and potentially revelatory. Debris, deposits, and deformations of various kinds—shards of glass and metal, indentations and punctures in vehicle components, dirt stains and blood spatters, fragments of fleshy tissue—have evidentiary significance, as do their precise locations and patterned configurations. They are clues, present and tangible, to how the events of the accident (absent, intangible) unfolded. "Evidence of marks, damage and debris suffered by or found on vehicles and highways will, if intelligently gathered and interpreted, assist in determining,

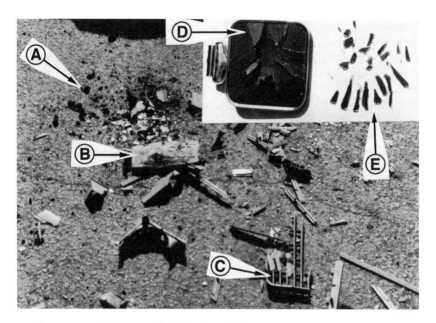

Accident-scene vehicle parts: (a) fender dirt, (b) license plate, (c) grille piece, (d) broken mirror, and (e) matching glass fragments. Source: R. W. Rivers, *Traffic Accident Investigators' Manual*, 2d edn. (1995). Courtesy of Charles C. Thomas, Publisher.

and in some cases conclusively establish, a vehicle's (or pedestrian's) direction of travel, placement and position and behavior during times of pre-collision, collision, and post-collision."[18] The field or ground containing the totality of collision imprints and residues, the terrain encompassing all the bits and pieces of bodies and machines, defines "the accident scene."

The accident scene is simultaneously effected and apprehended through devices and protocols of forensic mediation. Donald Van Kirk, author of *Vehicular Accident Investigation and Reconstruction*, has itemized the investigator's tools of the trade, his or her forensic hardware and software.[19] Plain paper is used to take handwritten notes, a portable audio recorder to take voice notes. A magnifying glass is used to inspect small objects and surfaces. A depth gauge or counter gauge is used to assess impact damage to vehicle interiors (as from heads or knees) and exteriors (as from signposts or tree trunks). Measuring tape and a measuring wheel are used to evaluate numerous details and distances within the scene. (Limpert: "Wherever possible, a parameter that *can* be measured *should* be measured and not estimated.")[20] Colored pens, graph paper, and a sighting compass are used to sketch the scene. Photographic (and, on oc-

Original caption: "A vehicle's path after collision may be traced from water or oil trails caused by the release of liquids from such parts as the radiator or transmission case. By following back along these trails from a vehicle's resting place, the approximate point of collision may often be established." Source: R. W. Rivers, *Traffic Accident Investigators' Manual*, 2d edn. (1995). Courtesy of Charles C. Thomas, Publisher.

casion, videographic) cameras, along with assorted lenses, filters, flash units, and film stocks, are used to shoot the scene. Writing and recording, magnifying and measuring, sketching and shooting: from this ensemble of observational instruments and representational techniques the accident scene is constituted as such. Which is to say that it is constituted as a distinct organization of material and symbolic space as well as a distinct object of technological and institutional perception, conception, and signification.

Noteworthy, in this connection, is the fact that accident-investigation and accident-reconstruction handbooks (most of which are so massive as to belie their status as "manuals" or "handbooks") devote considerable attention to methods of sketching, diagramming, and photographing—of visually rendering—the scene. Drawn in freehand and intended as a "pictorial supplement to the at-scene measurements," a field sketch includes "the results of the collision, roadway features, and significant objects which may have contributed to the collision."[21] By contrast, a scale diagram, or "after-collision situation map," is a more elaborate and meticulous endeavor, involving the use of drafting imple-

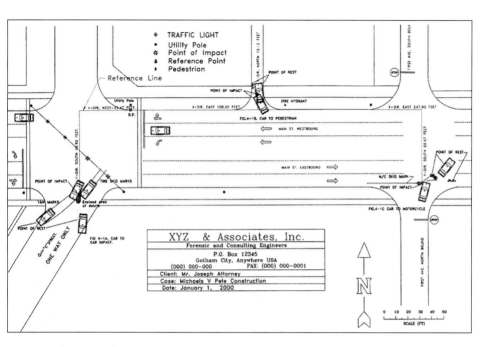

Accident-scene diagram. Source: Donald J. Van Kirk, *Vehicular Accident Investigation and Reconstruction* (2001). Courtesy of CRC Press.

ments (triangles, compasses, protractors, T-squares, parallel rulers, technical pens) coupled with the disciplined application of geometric and trigonometric principles. The scale diagram "should be a true and accurate representation of the collision scene after the traffic collision has stabilized," notes James Sneddon in Northwestern University Center for Public Safety's *Traffic Collision Investigation*. "It should contain only factual data." And plenty of them. "Tire marks, metal scars, and final rest positions of vehicles and bodies are depicted. Significant features such as parked vehicles, construction areas, and view obstructions are included. Roadway and roadside improvements such as pavement edges, lane markings, traffic signs and signals, fixed objects and sidewalks are also included."[22] The sketch, drawn at the scene as soon as feasible after a collision, is allied with note-taking and data gathering, with the exigencies of investigative fieldwork; it is rough, preliminary, and relatively swiftly executed, more handicraft than hard science. The situation map, on the other hand, is accomplished at a spatial and temporal remove from the scene, the result of cool, careful study and analysis; it is polished, finished, explanatory rather than documentary, a statement of scientific conclusion.

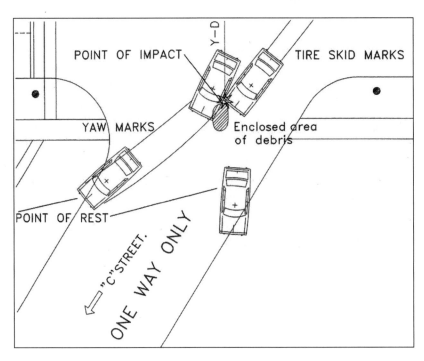

Detail of accident-scene diagram. Source: Donald J. Van Kirk, *Vehicular Accident Investigation and Reconstruction* (2001). Courtesy of CRC Press.

Belonging as it does to the realm of mechanical reproduction, photography—the prototypical modality of forensic vision and memory in technological modernity—differs categorically from sketching and diagramming.[23] Practically and epistemologically, the accident-scene photograph oscillates indeterminately between art and science, representation and indication, supplementarity and plenitude, appearance and empirical fact. Lynn Fricke and Kenneth Baker, in their chapter in *Traffic Collision Investigation*, assign photography a privileged role and a doubly preservative function. The photograph preserves, in a "permanent, accurate, unbiased" manner, what the investigator (consciously) observed at the scene and also, crucially, what he or she failed to (consciously) observe there. It serves as both "a record of observations" and "a reservoir of nondescript information," both a register of perceptions and a storehouse of stray details, a repository of previously unnoticed, potentially significant excesses and contingencies. By their very nature, "photos include much unnecessary data and may omit essential facts because the photographer was unaware of those facts when the photo was made. However, photos made

Original caption: "A single photograph should be taken showing all relevant detail and the relationships between things. Several photographs should also be taken from various angles to ensure all detail is captured." Source: R. W. Rivers, *Traffic Accident Investigators' Manual*, 2d edn. (1995). Courtesy of Charles C. Thomas, Publisher.

on the chance that they may prove useful do include an immense amount of data that would otherwise be unavailable. Moreover, photos made only to record an investigator's particular observations often also include a wealth of detail not noticed by the investigator at the time he made the photograph."[24]

The photographic image, in this formulation, is vitally important not only to the accident investigator, for whom it provides an extracorporeal means of ocular reduplication, but to the accident reconstructionist, from whom it, as a text to be read and reread, continually solicits close inspection. As a "permanent" storage medium, it allows a second look, and a third look — in fact, it admits an unlimited quantity of additional looks, invites a virtual infinity of forensic reviews. At the scene, during the work of investigation, the camera functions as an "unbiased" mechanism for the inscription of optical information, recording "facts about the situation more accurately, in greater detail, and more quickly than any observer possibly could."[25] After the scene, beyond the scene, during the work of reconstruction, the photograph, in suddenly bringing to light things initially unseen, in unexpectedly revealing what had been hidden or

Geometrizing the accident scene through photography. Source: R. W. Rivers, *Traffic Accident Investigators' Manual*, 2d edn. (1995). Courtesy of Charles C. Thomas, Publisher.

inconspicuous, becomes an instrument for piercing the veil of the ordinary visual world, for probing what Walter Benjamin famously (if not entirely aptly) called the "optical unconscious": those "physiognomic aspects" of phenomenal reality "which dwell in the smallest things" and are accessible only through photomechanical means; that normally obscured part of "nature that speaks to the camera rather than to the eye."[26]

Accident-scene photography, then, mobilizes a forensic visuality in several discrete yet interconnected moments during the investigation and reconstruction processes. In one instance, a forensic gaze is trained on the photographic image, searching its matrix of light and shadow for fugitive hints, for clues undetected the first time around, when the investigation was being conducted on the ground, "in the flesh." The activation of this gaze effectively affirms, in Susan Sontag's (Benjaminesque) words, "the belief that reality is hidden. And, being hidden, is something to be unveiled."[27] A paradigmatic expression of the forensic investment in the indexical trace, the forensic photographic gaze "presuppose[s] that photography provides a unique system of disclosures: that it shows us reality as we had *not* seen it before."[28] At other points, the photograph is forensically useful as a mnemonic device or, in the context of a legal proceeding, as an illustrative aid, a so-called demonstrative tool. Fricke and

Baker: "As a record of observations, photos serve (1) to recall later to an investigator's mind details of what he saw, and (2) to explain what the investigator saw to someone else, perhaps in court."[29]

There is another instance, too, that obeys a forensic visual imperative. Less overt and more abstract than those mentioned above, it is the moment that precedes any particular accident investigation or reconstruction and that conditions the possibility of all of them. It is the ur-moment that forms the investigative field of vision and ascribes importance (or insignificance) to certain elements and relations within that field. It is the anterior instance, notional but materially effective, that organizes the investigator's characteristic practices of looking and discriminating, that structures his or her peculiar faculty of attention and capacity for recognition and identification. It is the implicit foundation, the basic premise of investigative procedure at the scene, the rudimentary logic and discourse of evidentiary scrutiny and discovery. It is, finally, the institutionally embedded, ritually enacted framework of assumptions, the normative (and generative) scheme of designations and classifications, that regulates not only what is forensically seen and known but what there is to forensically see and know in the first place.[30]

Crucially, this complex of epistemic attitudes and perceptual dispositions is deeply—and antecedently—informed by techniques of mediated visualization. As Van Kirk explains, "It is both necessary and important that each investigator locates and records as much detail as possible when viewing the scene. Investigators must put themselves at the scene in their minds' eye, on the day of the accident. This scenario must play over and over in their minds, *as a movie or a videotape.*"[31] These remarks are reminiscent of both the film-theory concept of "mindscreen," which Bruce Kawin defines as "a visual (and at times aural) field that presents itself as the product of a mind, and that is often associated with systemic reflexivity, or self-consciousness," and the psychoanalyst of dreams, who, according to Sigmund Freud, "should picture the instrument which carries out our mental functions as resembling a compound microscope or a photographic apparatus, or something of the kind."[32] They are reminiscent, as well, of the narrator in the nineteenth-century detective-fiction pioneer Émile Gaboriau's "The Little Old Man of Batignolles," who surveys and subsequently remembers a crime scene with uncanny photographic precision: "I noticed these details in a moment, without a conscious act of will. My eyes became the lens of a camera, and the scene of the murder impressed itself on my mind, as if on a photographic plate, so precisely that even today I could draw the apartment of 'the little old man of Batignolles' from memory without forgetting anything."[33]

In Van Kirk's manual, the accident investigator is instructed to self-consciously analogize his or her own cognitive apparatus to something akin to a *kino-glaz*, a "film-eye" that fixes an image, as it were, without forgetting anything.[34] After sifting the evidence at the scene, the investigator must strive to imaginatively create and manipulate a compelling accident "scenario," to mentally record, rewind, and replay, "over and over" again if necessary, as on a movie or video screen, the events of the collision. His or her mind should resemble, in its inner workings, an advanced optical machine; visual media should be built into his or her modus operandi from the start. In this sense, the accident investigator, having internalized the dictates of a forensic regime of representation, is always already a disciplined picture-maker and picture-taker: a diagrammer even before the first sketch is drawn, a photographer (or cinematographer or videographer) even before the first shutter is snapped. And, by the same token, the accident scene is always already a technologically and institutionally mediated image.

The Trouble with Photography

Let us return for a moment to Fricke and Baker's appraisal of the utilities—and infelicities—of accident-scene photography. For that expert appraisal condenses and rehearses, in the register of the forensic, a number of familiar, not to mention fundamentally ambivalent, even contradictory, ideas and discourses concerning the ontological status of the photographic image. Consider, for example, the long-standing and recurrent question of photography's veracity, the nature of its realism. Although Fricke and Baker insist (rather conventionally) on the accuracy, impartiality, and mnemonic superiority of the photographic document—"Photos are certainly not subject to the loss of detail and uncertainties that [human] memory is"—they nevertheless acknowledge (also rather conventionally), as though guarding against the charge of epistemological naiveté, that "a photo may not be a completely 'true representation.'"[35] One reason why representational "distortions" occur, they contend, is that accident-scene photographers make mistakes, are prone to "easy errors." Some photographic improprieties are purely technical, such as those involving lens or focus or exposure or film development; others are essentially compositional, the result of "poor choices" regarding what to shoot (where to point the camera?) or how to shoot it (from what angle? what height? what distance?). "Technically perfect photos may not show what is required and they may even be misleading. In such cases, the photographer succeeds in operating his cam-

Original caption: "An investigator must know how to use his camera equipment properly. Improper camera settings can, for example, render a photograph useless." Source: R. W. Rivers, *Traffic Accident Investigators' Manual*, 2d edn. (1995). Courtesy of Charles C. Thomas, publisher.

era properly but fails in selecting the scene. If this happens, he may be accused of misrepresentation."[36]

In addition to errors of technique and failures of compositional choice, accident-scene photographers, in their quest to capture a "true representation," are susceptible to mistakes of moral judgment. "*Investigators can be led astray* by the obvious advantages of photography in collision investigation," warn Fricke and Baker. "These investigators are usually 'shutter-happy' camera fans who think photography is an easy way out of arduous observing and measuring."[37] Photography's "obvious advantages" can, on this view, be a dangerous lure. Indeed, in tempting the investigator to treat a camera as the fanatic does a fetish, the positive efficiencies of mechanical reproduction are reduced to magical distractions, inutile perversions. Technological ease becomes an enticement to waywardness, an inducement to indolence, while automaticity becomes a cheap substitute for scientific rigor.

Moreover, along with the charms and seductions intrinsic to the photographic apparatus, the accident investigator is obliged to resist the unseemly fascination aroused by "spectacular scenes": "Sometimes a badly injured person being extracted from a vehicle or firemen fighting a blaze is of compelling human interest. Newspaper photographers avidly seek opportunities to make such 'action' pictures. Yet they may be worthless for purposes of collision investigation. In fact, such pictures may not be permitted as evidence in court because they tend to bias the jury. Do not let yourself be carried away by such scenes."[38] Just as forensic-media practices are necessarily precarious, their close association with destruction, violence, and death constantly pressuring and threatening to destabilize their claims to dispassionate scientificity, so forensic-media practitioners, confronted with the collision's horrors and high dramas, are liable to be overwhelmed by their own emotions at the expense of attaining representational veracity. Their senses assaulted by too many stimuli, they might succumb to too much raw feeling. They could be diverted from their professional duties, enthralled by the visual attractions of bodies in action (whether the heroics of rescuers or the agonies of victims). The risk here is that in losing sight—quite literally—of what matters about the accident scene, of what therein is forensically pertinent and proper, investigators unwittingly vitiate, potentially to the point of total invalidation, the photographic evidence they labor to produce. Thus, as Fricke and Baker suggest, the boundary between forensic photography and yellow photojournalism, between clinical realism and tabloid sensationalism, between a veridical image and a prejudicial one, is thin, permeable, and ever in need of vigilant patrol.

The trouble with accident-scene photography, however, is not limited to errors of camera operation and of shot selection and of pictorial discretion, to failures technical, compositional, and moral. No, the trouble with accident-scene photography—as with all photography—runs deeper still. For apart from the difficulties attending the authoritative *production* of accident-scene photographs, there are those attending their authoritative *consumption*. Instead of the complexities of *executing* such photographs, these latter difficulties involve the perplexities of *interpreting* them. Such hermeneutic perplexities frequently arise from the encounter with the singularities and surpluses of the photographic image, from the attempt to grapple with the sheer volume of visual material that image (never fully) contains.

Writing on the use of photography in the biological sciences, Michael Lynch points out that photographs sometimes manifest "too much reality" for scientists to smoothly handle:

> The circumstantial sensitivity and singularity of a photograph presents a problem for any effort to conceptualize the subject and represent abstract

features. Although a photograph is set up through complex arrangements of the pose, lighting conditions, exposure, frame, and focus, the resultant picture may seem indifferent to what it "captures." It does not reproduce what an observer originally experiences, since it exposes an entire field of light, often including unseen, unanticipated, and unwanted visible configurations. In different circumstances such features can be treated as distortions of a scene's original features, or as surplus details that contribute to a heightened sense of the scene's reality. Ultimately, such surplus details can transform the very meaning of the "original" or "unmediated" scene.[39]

Recall that, even as they extol the epistemological virtues of photography, Fricke and Baker are moved to concede that "photos include much unnecessary data." This concession would seem to warrant—but, curiously and perhaps tellingly, does not receive—further comment, as it raises questions about what exactly is entailed, and what ultimately is at stake, in deciphering or decoding an accident-scene photograph. What is the difference between "necessary" and "unnecessary" data? How is such a determination made, and how is such a distinction maintained, in forensic theory and practice?

The problem of photographic excess, the question of which aspects of a given image are essential and which accidental, of "message" versus "noise," is as old as the medium itself. Discussing the nexus between early photography and the emergence of the detective story, Robert Ray notes that

> photography could boast a pseudoscientific basis and play into the late-eighteenth- and early-nineteenth-century rage for classification. But while the new technology seemed the ideal means of gathering the empirical data required by any system, almost immediately the first photographers noticed something going wrong. One historian cites [the inventor and photography pioneer] Fox Talbot's surprise at what he found: "And that was just the trouble: fascinating irrelevancy. 'Sometimes inscriptions and dates are found upon buildings, or printed placards most irrelevant, are discovered upon their ways: sometimes a distant sundial is seen, and upon it—unconsciously recorded—the hour of the day at which the view was taken.' To judge from his commentaries, Fox Talbot enjoyed such incidentals. At the same time, though, they were troublesome, for they meant that the instrument was only partially under control, recording disinterestedly in despite of its operator's intentions."[40]

From the beginning, then, the camera's ability to "disinterestedly" record and store optical information has carried a double valence. On the one hand, photography promises to automate and thereby improve methods of data collection and systematic classification, to vastly expand and refine knowledge in the name of objective science. On the other hand, the unintended appearance, here and there in the photographic image, of tantalizing contingencies, of "fascinating irrelevancies," threatens not only to shatter the illusion of technological mastery but also to corrupt the integrity and undermine the sovereignty of scientific vision, to impede the ambitions of analytical reason by enabling an illicit, or at least an unauthorized, mode of subjective enjoyment.

Like William Henry Fox Talbot in the prior century, both Walter Benjamin and Roland Barthes (as well as the Surrealists who influenced them) took a sort of ecstatic pleasure in photographic "incidentals."[41] Captivated by the random traces of reality that "seared through the image-character of the photograph," Benjamin wrote of his "unruly desire to know," his "irresistible compulsion to search" the photographic image for its "tiny spark of contingency."[42] With these words, Benjamin, himself an avid reader of detective fiction, might as well have been describing the forensic photographic impulse.[43]

Barthes, for his part, argued in an essay published in 1970 that some photographs harbor a "third meaning," an "obtuse meaning" that surpasses both the primary informational and the secondary symbolic meanings.[44] Composed of "signifying accidents," the third meaning "appears to extend outside culture, knowledge, information."[45] As "the one 'too many,' the supplement that [one's] intellection cannot succeed in absorbing," it charts "the *passage* from language to *signifiance*"—the latter a term, as Stephen Health observes, linked to the "climactic," dissipative, "radically violent" pleasure of *jouissance*.[46]

Barthes returned to the theme of photographic excess in 1980, with the publication of *Camera Lucida*. In this book, the photographic image's overabundance is recast as acute rather than obtuse, no longer a meaning (tertiary or otherwise) but a feeling, a bodily sensation, keen and seemingly immediate: a *punctum*.[47] As against the culturally coded *studium*, which elicits low-intensity ("vague, slippery, irresponsible") interest in a photograph's normative content (its ethical, political, or historical significance), the uncoded, extracultural punctum "shoots out of [a photograph] like an arrow, and pierces" the beholder, virtually wounding him.[48] As Barthes says, the punctum, "when it happens to be there"—for it is so only rarely and, as the very mark of contingency, only by chance—is a "sting, speck, cut, little hole—and also a cast of the dice. . . . that accident which pricks me (but also bruises me, is poignant to me)."[49] Whereas the docile, conventional studium may bring ordinary pleasure (*plaisir*), the intractable, unpredictable punctum delivers injurious enjoyment (*jouissance*).

Now, there is no guarantee that an accident investigator or reconstructionist, scanning an accident-scene photograph for its forensic studium, will not be pricked by an irrelevant punctum; no reason to suppose that he or she will be immune to the allure of "signifying accidents." To be sure, the possibility always exists that even the most intently focused investigator or reconstructionist will become distracted by pictorial marginalia, enchanted by the minor detail, mesmerized by the intellectually unassimilable item — or, alternatively, will be unable to distinguish useful information from worthless non-information, the forensically meaningful from the forensically null. Such perceptually and semiotically disruptive possibilities have haunted photography, including forensic photography, since its inception. "While the longing for strictly objective, and therefore *exact*, representation had motivated photography's invention, photographs produced precisely the opposite effect — a mute ambiguity that invited subjective reverie," writes Ray in reference to nineteenth-century photography.[50] "By dramatically increasing the available amount of particularized information, photography . . . ensured that in every context where it intervened, distinguishing the significant from the insignificant would become treacherous."[51] Representational ambiguity and hermeneutical treachery: are these not precisely the implications, latent or repressed, of Fricke and Baker's "much unnecessary data" remark?

Yet, according to these same authors, it will be recalled, photography's "too muchness" — its visual surfeit, its optical profusion — also contains a kernel of forensic opportunity: "Photos made on the chance that they may prove useful do include an immense amount of data that would otherwise be unavailable. Moreover, photos made only to record an investigator's particular observations often also include a wealth of detail not noticed by the investigator at the time he made the photograph." The specter of contingency is here tied to the production of photographic evidence in a couple of respects. First, a certain calculated embrace of fortuitousness, a certain strategic aleatory or stochastic practice, is prescribed at the level of investigative fieldwork: because the camera "sees" indifferently and without intentionality, the accident-scene photographer is encouraged to snap a few random photos, to take a few pictures of nothing in particular, "on the chance that" one or more of them will "prove useful." (Sontag names this "almost superstitious" technique, routinely employed by photographers of every stripe, "the scattershot method.")[52] Second, the photographic image, "swarming," in Ray's words, "with accidental details unnoticed at the time of shooting," makes possible the "type of surprise" Barthes calls "the *trouvaille* or lucky find": the chance discovery, the unexpected revelation.[53] (Conversely, it also makes possible the unforeseen and unbidden enigma, the inscrutable forensic punctum — exactly the issue dramatized in Michelangelo

Film still from *Blow-Up* (1966), directed by Michelangelo Antonioni.

Antonioni's *Blow-Up*.) In this formulation, photography's supplementarity—all those "unconsciously recorded" singularities, all those incidentals that "sear through" or "shoot out of" the image surface—is precisely the precondition for mediated forensic serendipity. (Conversely, it also is the precondition for mediated forensic puzzlement, for artifactual "detective anguish"—again, *Blow-Up* is the salient reference.)[54]

Back to (the Specter of) Zadig

That the forensic photographic gaze welcomes and readily incorporates, rather than refuses, the happy accident comes as no great surprise. In fact, genealogically, accident-scene photography's special allowance for the vagaries of fortune could hardly be more fitting. As both Carlo Ginzburg and Jean-Michel Rabaté (following Régis Messac) have suggested, the relation between forensic detection and serendipitous discovery has a long history: it is truly the stuff of legend.[55]

In the folktale, probably of ancient Persian origin, known in English as "The Three Princes of Serendip," the sons of King Giaffer are able to accurately de-

Frontispiece of Michele Tramezzino's *Peregrinaggio di tre giovani figliuoli del re di Serendippo* (1557).

scribe the peculiar traits of a missing camel without having actually seen it.[56] The princes accomplish this astonishing and uncanny feat by inspecting the animal's barely noticeable tracks and traces, its residual stains and imprints. (Using the same method, they also accurately describe the peculiar traits of the camel's rider.) Ginzburg discerns in this tale, famously adapted in 1747 by Voltaire in chapter 3 of *Zadig*, both "an echo, though dim and distorted, of the knowledge accumulated by [prehistoric] hunters" and a striking anticipation of the modern detective story:

> Man has been a hunter for thousands of years. In the course of countless chases he learned to reconstruct the shapes and movements of his invisible prey from tracks on the ground, broken branches, excrement, tufts of hair, entangled feathers, stagnating odors. He learned to sniff out, record, interpret, and classify such infinitesimal traces as trails of spittle.

He learned how to execute complex mental operations with lightning speed, in the depth of a forest or in a prairie with its hidden dangers. . . .

Obviously, the three brothers are repositories of some sort of venatic lore, even if they are not necessarily hunters. This knowledge is characterized by the ability to construct from apparently insignificant experimental data a complex reality that could not be experienced directly.[57]

In Voltaire's adaptation, the three princes have been refashioned as one character, the novel's namesake (a heterodox Babylonian philosopher), and the one camel has been refashioned as two animals, a bitch and a horse. But the emphasis on the application of neo-venatic—indeed, proto-forensic—knowledge remains essentially the same. At a crucial point in the story, Zadig recounts

> the mental process which had enabled him to sketch the portrait of two animals he had never seen: "I saw on the sand the tracks of an animal, and I easily judged that they were those of a little dog. Long, shallow furrows imprinted on little rises in the sand between the tracks of the paws informed me that it was a bitch whose dugs were hanging down, and that therefore she had had puppies a few days before." These lines, and those which followed, were the embryo of the mystery novel. They inspired [Edgar Allan] Poe, Gaboriau, and [Arthur] Conan Doyle—the first two directly, the third perhaps indirectly.[58]

Seven years after Voltaire's retelling of "The Three Princes of Serendip," Horace Walpole, originator of that most haunting of literary genres, the gothic novel, coined the term *serendipity* to designate the making of unsought discoveries "by accidents and sagacity."[59]

notes

introduction. ACCIDENTS AND FORENSICS

1. Until 1949, Edwards Air Force Base was known as Muroc Army Air Field.
2. On Stapp's rocket-sled experiments at Edwards, see Stapp, "Human Exposures to Linear Deceleration"; Spark, "The Fastest Man on Earth"; and Stapp, "Human Tolerance to Deceleration."
3. My discussion of the legend of Murphy's Law draws on Spark, "The Fastest Man on Earth."
4. This uncommon variant is notable as the only one that does not imply the notion of error or failure.
5. "Crash of Flight 447."
6. Downie and Wald, "More Bodies Recovered Near Site of Plane Crash," www.nytimes.com/2009/06/08/world/americas/08plane.html.
7. Clark, "Bodies from 2009 Air France Crash Are Found," http://www.nytimes.com/2011/05/17/world/europe/04airfrance.html.
8. Clark, "Silence Still from Resting Place of Air France Recorder," http://query.nytimes.com/gst/fullpage.html?res=9B02E7D61E3DF933A05755C0A96F9C8B63.
9. Clark, "Bodies from 2009 Air France Crash Are Found."
10. See Clark, "Bodies from 2009 Air France Crash Are Found"; and Hylton, "The Deepest End." REMUS is an acronym for Remote Environmental Monitoring Unit System.
11. Clark, "Second Black Box Found in Air France Crash," www.nytimes.com/2011/05/04/world/europe/04airfrance.html.
12. Clark, "Data Recovered from Air France Flight Recorders," www.nytimes.com/2011/05/17/world/europe/17airfrance.html.
13. Ibid.
14. Bureau d'Enquêtes et d'Analyses pour la Sécurité de l'Aviation Civile, *Final Report on the Accident on 1st June 2009 to the Airbus A330-203*, 203.
15. I borrow the term *protocols* from Lisa Gitelman, who writes that protocols "include a vast clutter of normative rules and default conditions, which gather and adhere like a nebulous array around a technological nucleus. Protocols express a huge variety of social, economic, and material relationships" (*Always Already New*, 7).
16. Calinescu, *Five Faces of Modernity*, 41.

17. G. W. F. Hegel, quoted in Marquard, *In Defense of the Accidental*, 109.
18. Witmore, *Culture of Accidents*, 43.
19. Ibid., 2.
20. Ibid.
21. Piaget, *The Child's Conception of Physical Causality*, 117–18.
22. Lévy-Bruhl, *Primitive Mentality*, 43.
23. See Evans-Pritchard, "Lévy-Bruhl's Theory of Primitive Mentality."
24. Evans-Pritchard, *Witchcraft, Oracles and Magic among the Azande*, 72.
25. Ibid., 70.
26. Judith Green, *Risk and Misfortune*. See also Judith Green, "Accidents."
27. Green, *Risk and Misfortune*, 58.
28. Paz, *Conjunctions and Disjunctions*, 112.
29. On the "supplement," see Derrida, *Of Grammatology*. On the "residual category," see Judith Green, *Risk and Misfortune*; and Judith Green, "Accidents."
30. Bauman, *Modernity and Ambivalence*, 1.
31. Quoted in Marx, *The Machine in the Garden*, 170.
32. Henry Adams, *The Education of Henry Adams*, 495.
33. Singer, *Melodrama and Modernity*, 70–71.
34. Beck, *Risk Society*, 21, emphasis omitted.
35. Giddens, *The Consequences of Modernity*, 111. See also Giddens, *Modernity and Self-Identity*.
36. Ibid.
37. Ewald, "Two Infinities of Risk," 226.
38. Ibid.
39. Ibid. See also Ewald, "Insurance and Risk."
40. Foucault, *Politics, Philosophy, Culture*, 257. See also Foucault, "Polemics, Politics, and Problemizations."
41. Foucault, *Power/Knowledge*, 131.
42. Cooter and Luckin, "Accidents in History," 3. See also Hacking, *The Emergence of Probability*; Hacking, *The Taming of Chance*; and Hacking, "How Should We Do the History of Statistics?"
43. Cooter and Luckin, "Accidents in History," 4.
44. Packer, "Disciplining Mobility," 149.
45. Koolhaas, "'Life in the Metropolis' or 'The Culture of Congestion,'" 324–25.
46. Hobbes, *Leviathan*, 63.
47. Ibid., 64.
48. Hume, *The Natural History of Religion*, 140.
49. Ibid., 143.
50. Barthes, *The Grain of the Voice*, 115.
51. Butler, *Prose Observations*, 33.
52. Foucault, *Discipline and Punish*, 23.
53. With, *Railroad Accidents*, 11.
54. Ibid.
55. Ibid., 124–25.

56. Barstow, preface, vi.

57. Nietzsche, *On the Genealogy of Morality*, 104. Nietzsche's critique of "the causal instinct" in *Twilight of the Idols* is pertinent here: "To derive something unknown from something familiar relieves, comforts, and satisfies, besides giving a feeling of power. With the unknown, one is confronted with danger, discomfort, and care; the first instinct is to abolish these painful states. . . . The causal instinct is thus conditional upon, and excited by, the feeling of fear. . . . Thus one searches not only for some kind of explanation to serve as a cause, but for a particularly selected and preferred kind of explanation—that which has most quickly and most frequently abolished the feeling of the strange, new, and hitherto unexperienced: the *most habitual explanations*" (*The Portable Nietzsche*, 497).

58. Lévi-Strauss, *Structural Anthropology*, 209.

59. Žižek, *The Sublime Object of Ideology*, 3.

60. Doane, *The Emergence of Cinematic Time*, 10.

61. Ibid., 33–34.

62. Krauss, *The Originality of the Avant-Garde and Other Modernist Myths*, 218.

63. Beckman, *Crash*, 131. On the role of "virtual witnessing" in the history of science, see Shapin and Schaffer, *Leviathan and the Air-Pump*, 60–63.

64. Foucault, *The Birth of the Clinic*, 90, emphasis omitted.

65. Ibid., emphasis omitted.

66. Ibid.

67. Ginzburg, *Clues, Myths, and the Historical Method*, 103.

68. Ibid.

69. Ibid.

70. The full-length essay was first published in English, with a translation by Anna Davin, as "Morelli, Freud and Sherlock Holmes: Clues and Scientific Method," *History Workshop: A Journal of Socialist Historians* 9 (spring 1980): 5–36. A few years later, it was reprinted under the title "Clues: Morelli, Freud, and Sherlock Holmes," in *The Sign of Three: Dupin, Holmes, Peirce*, ed. Umberto Eco and Thomas A. Sebeok (Bloomington: Indiana University Press, 1983), 81–118. In 1989, the essay was retranslated and retitled for inclusion in Ginzburg's book *Clues, Myths, and the Historical Method*.

71. Nisbet, *History of the Idea of Progress*, 4–5, emphasis omitted.

72. Giedion, *Mechanization Takes Command*, 31. Lewis Mumford makes essentially the same point: "With the rapid improvement of machines, the vague eighteenth century doctrine [of progress] received new confirmation in the nineteenth century. The laws of progress became self-evident: were not new machines being invented every year? Were they not transformed by successive modifications? Did not chimneys draw better, were not houses warmer, had not railroads been invented?" (*Technics and Civilization*, 182).

73. Mazlish and Marx, "Introduction," 1.

74. Adas, *Machines as the Measure of Men*.

75. Ibid., 379–80.

76. Giedion, *Mechanization Takes Command*, 715.

77. Schivelbusch, *The Railway Journey*, 131.

78. Doane, "Information, Crisis, Catastrophe," 231.

79. Žižek, *The Sublime Object of Ideology*, 70.

80. Marx, "The Domination of Nature and the Redefinition of Progress," 210. Nisbet writes, "There is . . . good ground for supposing that when the identity of [the twentieth] century is eventually fixed by historians, not faith but abandonment of faith in the idea of progress will be one of the major attributes" (*History of the Idea of Progress*, 317).

81. Adorno, "Progress," 94.

82. Virilio, *Politics of the Very Worst*, 92. See also Arendt: "Progress and catastrophe are the opposite faces of the same coin" (quoted in Virilio, *Unknown Quantity*, 40).

chapter one. ENGINEERING DETECTIVES

Thomas Tredgold is quoted in Mitcham and Schatzberg, "Defining Technology and the Engineering Sciences," 41.

1. Ward, "What Is a Forensic Engineer?," 1.
2. Noon, *Forensic Engineering Investigation*, doi: http://dx.doi.org/10.1201/9781420041415.
3. *Oxford English Dictionary Online*, s.v. "forum, n.," www.oed.com.
4. Keenan and Weizman, "Mengele's Skull," 64. For an extended treatment of this topic, see Keenan and Weizman, *Mengele's Skull*.
5. Carper, foreword, iii.
6. Ibid.
7. Petroski, *To Engineer Is Human*, viii.
8. Ibid., 163.
9. Ibid., 97. See also Ihde, *Technology and the Lifeworld*.
10. Godfrey, *Engineering Failures and Their Lessons*, v and passim.
11. Ibid., vii.
12. Ibid., viii.
13. Ibid., x.
14. Ibid., xiii.
15. Manby, *Minutes of Proceedings of the Institution of Civil Engineers*, 455.
16. Petroski, *To Engineer Is Human*, 9.
17. Ibid., 10.
18. Ibid., 231.
19. Ibid., 28.
20. Ibid., 105.
21. George H. Thomson, "American Bridge Failures," 294. Note that Thomson here cites Thomas Tredgold's famous definition of "engineering," which serves as an epigraph to the present chapter.
22. Godfrey, *Engineering Failures and Their Lessons*, xi.
23. Lossier's book was originally published in French in 1952; the English edition translates the second French edition, which appeared in 1955. The Jacob Feld quote is from R. F. Legget's preface to the English edition.
24. Gordon, *Structures: or Why Things Don't Fall Down*, 324.
25. Ward, "What Is a Forensic Engineer?," 1.
26. Carper, "What Is Forensic Engineering?," 5.

27. Shackelford, "Failure Analysis," doi: http://dx.doi.org/10.1201/9781420039870.
28. Petroski, "The Success of Failure," 323.
29. "The key feature of the forensic gaze," according to Mariana Valverde, is "the close attention to the physical traces left not only by criminal activity but by everyday activity on people's bodies and clothes, on floors, walls, gardens and objects" (*Law and Order*, 83).
30. Foucault, *The Birth of the Clinic*, 146.
31. Ibid., 135–36. For an excellent discussion of "the corpse as spectacular dis-membered specimen," see chapter 4 of Robert D. Romanyshyn's *Technology as Symptom and Dream*.
32. *Oxford English Dictionary Online*, s.v. "autopsy, n.," http://www.oed.com.
33. Foucault, *The Birth of the Clinic*, 166.
34. Ibid., 141.
35. Sawday, *The Body Emblazoned*, 1.
36. It is useful here to keep in mind Allan Sekula's clarification of the terminological distinction between "criminology" and "criminalistics": "Criminology hunted 'the' criminal body. Criminalistics hunted 'this' or 'that' criminal body" ("The Body and the Archive," 18).
37. Thorwald, *Crime and Science*, 234–35.
38. Quoted in ibid., 282.
39. Ronald R. Thomas, *Detective Fiction and the Rise of Forensic Science*, 4.
40. Ibid., 3.
41. Ibid., 4
42. Bloch, *Literary Essays*, 219.
43. Ibid., 225.
44. Ibid., 213
45. Ibid.
46. Ibid., 220
47. Ibid., 215
48. Ibid., 213
49. Porter, "Backward Construction and the Art of Suspense," 334.
50. Ronald R. Thomas, *Detective Fiction and the Rise of Forensic Science*, 6.
51. Ibid., 17.
52. Ibid.
53. Bloch, *Literary Essays*, 219.
54. On the idea of forensic "imaginary replay," see Valverde, *Law and Order*, 83.
55. Porter, "Backward Construction and the Art of Suspense," 334.
56. Petroski, *Design Paradigms*, 12.
57. To be clear, my argument here is that the seeds for what would eventually be called "forensic engineering" were sown in the nineteenth century, even though the designation itself does not appear to have gained currency until the second half of the twentieth.
58. On the idea of modern technology run amok, see Winner, *Autonomous Technology*.
59. Hunter, *Steamboats on the Western Rivers*, 271.
60. Ibid., 282.
61. Schivelbusch, *The Railway Journey*, 78.
62. Aldrich, *Death Rode the Rails*, 2.

63. Reed, *Train Wrecks*, 25.

64. Singer, *Melodrama and Modernity*, 62.

65. Hunter, *Steamboats on the Western Rivers*, 289.

66. Ibid., 295.

67. The title of this section alludes to cultural studies' commitment to what Stuart Hall, following Karl Marx, has termed "the necessary detour through theory." See Stuart Hall, "Cultural Studies and Its Theoretical Legacies." See also Grossberg, *Bringing It All Back Home*, especially the chapter "Cultural Studies: What's in a Name? (One More Time)," 245–71.

68. Virilio, *Unknown Quantity*, 23, 24.

69. Virilio, *Politics of the Very Worst*, 89.

70. Virilio, "The Primal Accident," 212.

71. Ibid.

72. Virilio, *Politics of the Very Worst*, 89.

73. Virilio quotes Nicholas of Cusa, in *Open Sky*, 17.

74. Virilio, "The Primal Accident," 211.

75. The interpretative issue here is whether Virilio is making an essentially *historical* or an essentially *philosophical* claim. Have the circumstances of technological modernity rendered outdated the Aristotelian view of substance and accident? Or, by contrast, have they merely served to highlight the fact that that view was never sound to begin with? Peter van Wyck reads Virilio as making the former claim (*Signs of Danger*, 12–13). While I, too, am inclined to this interpretation, the ambiguity of Virilio's formulation is worth noting.

76. Virilio, *The Original Accident* (this book includes different translations of the essays from *Unknown Quantity*); Virilio, "The Primal Accident."

77. Virilio, "The Primal Accident," 212.

78. Ibid.

79. Hamilton, *Accident*, 16.

80. Aristotle, *Metaphysica*, book 5, chap. 30, 1025a.

81. Meyer, "Aristotle, Teleology, and Reduction," 795.

82. Ibid, 802.

83. I am alluding here to the Epicurean notion of atomic unpredictability/quantum indeterminacy, which Lucretius called "*clinamen*." See Michel Serres's essay "Lucretius: Science and Religion," in Serres, *Hermes*; and Greenblatt, *The Swerve*.

84. Michael Witmore, *Culture of Accidents*, 28.

85. Ibid.

86. Aristotle, *Metaphysica*, book 6, chap. 2, 1026a.

87. Ibid., book 6, chap. 4, 1027b.

88. Ibid., book 6, chap. 2, 1026b.

89. Ibid.

90. Quoted in Witmore, *Culture of Accidents*, 37.

91. Aristotle, *Aristotle's Physics*, 354.

92. Witmore, *Culture of Accidents*, 38.

93. Ibid., 39.

94. Ibid., 11.

95. Brunvand, *The Vanishing Hitchhiker*, 12. See also the entries on "Accidents" and "Automobile Accidents" in Brunvand, *Encyclopedia of Urban Legends*.

96. Schivelbusch, *The Railway Journey*, 153.

97. Schivelbusch puts it this way: "The higher the degree of technical intensification (pressure, tension, velocity, etc.) of a piece of machinery, the more thorough-going was its destruction in the case of dysfunction" (*The Railway Journey*, 131). Virilio makes essentially the same point: "No gain without a corresponding loss. If to invent the substance is, indirectly, to invent the accident, then the more powerful and efficient the invention, the more dramatic the accident" (*Unknown Quantity*, 85).

98. Schivelbusch, *The Railway Journey*, 162. Schivelbusch here anticipates, in a Freudian register, Edward Tenner's (non-Freudian) thesis in *Why Things Bite Back*.

99. Goodman, *Shifting the Blame*, 69.

100. Ibid., 4.

101. Horwitz, *The Transformation of American Law, 1780–1860*, 88.

102. Goodman, *Shifting the Blame*, 74–75.

103. Both Goodman and Horwitz consider the complex legal and philosophical question of proximate versus remote causation, and both reference Nicholas St. John Green's fascinating *American Law Review* article "Proximate and Remote Cause," published in 1870 and reprinted in his *Essays and Notes on the Law of Tort and Crime*, 1–17. On the legal concept of "ordinary care," see Goodman's introduction to *Shifting the Blame*.

104. Goodman, *Shifting the Blame*, 37.

105. Paz, *Conjunctions and Disjunctions*, 111.

106. *Oxford English Dictionary Online*, s.v. "accident, n.," www.oed.com.

107. Schivelbusch, *The Railway Journey*, 131.

108. *Oxford English Dictionary Online*, s.v. "catastrophe, n.," www.oed.com.

109. Doane, "Information, Crisis, Catastrophe," 229.

110. Brockmann, *Twisted Rails, Sunken Ships*, 8.

111. Quoted in Brockmann, *Twisted Rails, Sunken Ships*, 8–9.

112. Charles Francis Adams Jr., *Notes on Railroad Accidents*, 86.

113. Ibid., 85–86.

114. Ibid., 86.

115. Stanley Hall, *Railway Detectives*, 18.

116. Ibid., 20.

117. *Oxford English Dictionary Online*, s.v. "detective, n.," www.oed.com.

118. Gross, *Criminal Investigation*, 859. Gross adds that, even though he uses "boiler explosions as a type," all of the accidents he lists should be investigated using the same method: "The *same method* of inquiry will apply, *mutatis mutandis*, to all serious catastrophes of the nature now under consideration" (863).

119. Ibid., 860.

120. Ibid., 865.

121. Ibid., 863.

122. Ibid., 859.

123. On the technological sublime, see Nye, *American Technological Sublime*. On the

Corliss engine and what the author calls "the aesthetics of machinery," see Kasson, *Civilizing the Machine*. I discuss the technological sublime in the next chapter.

124. See Mumford, *Technics and Civilization*.

125. Brockmann, *Twisted Rails, Sunken Ships*, 14.

126. Brockmann writes, "The role of the 'disinterested' expert grew because scientific expertise—once centered in the Franklin Institute and its loose association of scientist volunteers in the Committee [on] Science and the Arts—greatly expanded through the Civil War and became more widely dispersed. For example, so many people became civil engineers that a professional society, the American Society of Civil Engineers, was founded in 1852 and began to play a role in investigations" (*Twisted Rails, Sunken Ships*, 176).

127. Charles Francis Adams Jr., *Notes on Railroad Accidents*, 2.

128. Bloch, *Literary Essays*, 307.

129. Ibid., 305.

130. Ibid., 307.

131. Ibid., 310.

132. Ibid., 312.

133. Baudrillard, *Symbolic Exchange and Death*, 161.

134. Tichi, *Shifting Gears*, 105.

135. Keenan and Weizman, "Mengele's Skull," 64.

136. Noon, *Forensic Engineering Investigation*, n.p.

137. George H. Thomson, "American Bridge Failures," 294.

138. Ronald R. Thomas, *Detective Fiction and the Rise of Forensic Science*, 4.

chapter two. TRACINGS

Étienne-Jules Marey is quoted in Brain, "Representation on the Line," 157.

1. Babbage, *Passages from the Life of a Philosopher*, 245.

2. Ibid., 239.

3. Hyman, *Charles Babbage*, 160.

4. Babbage, *Passages from the Life of a Philosopher*, 248.

5. Hyman, *Charles Babbage*, 160.

6. Hankins and Silverman, *Instruments and the Imagination*, 128.

7. Ibid., 135. For a fascinating take on Scott de Martinville's phonautograph, see "Machines to Hear for Them," the first chapter in Sterne, *The Audible Past*.

8. Quoted in Braun, *Picturing Time*, 15, 12–13.

9. Quoted in Rabinbach, *The Human Motor*, 95.

10. Quoted in Braun, *Picturing Time*, 40.

11. Quoted in Rabinbach, *The Human Motor*, 95.

12. Brain, "Representation on the Line," 156.

13. Quoted in Rabinbach, *The Human Motor*, 94.

14. Ludwig Wittgenstein invoked the term *family resemblances* to specify "a complicated network of similarities overlapping and criss-crossing: sometimes overall similarities, sometimes similarities of detail" (*Philosophical Investigations*, 32). His reference was to different kinds of (language) game; mine is to different kinds of technology.

15. Babbage, *Passages from the Life of a Philosopher*, 245.
16. Quoted in Doane, *The Emergence of Cinematic Time*, 48, emphasis added.
17. Quoted in Hankins and Silverman, *Instruments and the Imagination*, 135.
18. Ibid., 139.
19. Peters, "Helmholtz, Edison, and Sound History," 178.
20. Brain, "Representation on the Line," 166.
21. Reed, *Train Wrecks*, 25.
22. Schivelbusch, *The Railway Journey*, 78.
23. Cooter, "The Moment of the Accident," 112.
24. Babbage, *Passages from the Life of a Philosopher*, 241.
25. Ibid., 237.
26. Ibid., 238, 245.
27. "The result of my experiments convinced me that the broad gauge was most convenient and safest for the public" (Babbage, *Passages from the Life of a Philosopher*, 239). For more on Babbage's take on the question of track gauge, see pages 243 and 249–50 of his autobiography. Incidentally, Babbage and the other broad-gauge supporters ended up losing the battle, as narrow gauge became the British standard in 1846 through an Act of Parliament.
28. Ibid., 234.
29. Ibid., 235.
30. Hyman, *Charles Babbage*, 144.
31. Babbage, *Passages from the Life of a Philosopher*, 235.
32. Marx, *The Machine in the Garden*, 195.
33. Kasson, *Civilizing the Machine*; Nye, *American Technological Sublime*. Although Marx, Kasson, and Nye all focus on the American context, plenty in their analyses is applicable to the concurrent Western European context.
34. Quoted, respectively, in Eagleton, *The Ideology of the Aesthetic*, 52, and in Kasson, *Civilizing the Machine*, 166.
35. Quoted in Kasson, *Civilizing the Machine*, 166.
36. Quoted in Nye, *American Technological Sublime*, 9.
37. Quoted in ibid., 7.
38. Quoted in Nye, *American Technological Sublime*, 7.
39. Kasson, *Civilizing the Machine*, 166–67.
40. Ibid., 172.
41. Marx, *The Machine in the Garden*, 197.
42. On "structure of feeling," see Williams, *The Long Revolution*, 48–49.
43. Adorno, "Progress," 89.
44. Jünger, "On Danger," 370.
45. Schnapp, "Crash (Speed as Engine of Individuation)," 14.
46. Ibid., 3.
47. Schivelbusch, *The Railway Journey*, 162.
48. Giedion, *Mechanization Takes Command*, 30.
49. Ibid., 31.
50. Trachtenberg, foreword, xiii.

51. I discuss Her Majesty's Railway Inspectorate in the "From Coroners' Juries to Railway Detectives" section of chapter 1.

52. On the history of the seismograph, see Dewey and Byerly, "The Early History of Seismometry (to 1900)." On the history of the lie detector, see Trovillo, "A History of Lie Detection." See also Ronald R. Thomas's superb literary and cultural analysis of the lie detector in *Detective Fiction and the Rise of Forensic Science*.

53. Babbage, *Passages from the Life of a Philosopher*, 249.

54. Ibid., 245.

55. Babbage proved remarkably prescient in this regard, predicting that his apparatus probably would "be allowed to go to sleep for years, until some official person, casually hearing of it, or perhaps re-inventing it, shall have *interest* with the higher powers to get it quietly adopted as his own invention" (ibid., 245).

chapter three. BLACK BOXES

1. Wiener, *Cybernetics*, xi, n. 1.

2. See especially Serres, *The Parasite*; and Latour, *Science in Action*. See also Latour, *Pandora's Hope*; and Serres, *The Five Senses*.

3. Latour, *Science in Action*, 7.

4. Carl Bennett and the Silent Era Company's "Progressive Silent Film List" includes the full synopsis of *The Black Box* as printed in the film company's promotional literature. See Bennett, *The Black Box*, www.silentera.com/PSFL/data/B/BlackBox1915.html.

5. Oppenheim, *The Black Box*, 58–59.

6. Ibid., 138.

7. Ibid., 3.

8. Ibid., 57–58.

9. Kittler, *Gramophone, Film, Typewriter*, 255.

10. Rosenheim, *The Cryptographic Imagination*, 3.

11. Ibid., 6.

12. See Goldhurst, "A Literary Source for O'Neill's 'In the Zone'"; and Tamár, "Possible Sources for Two O'Neill One-Acts."

13. The 1902 edition of Eastman Kodak's promotional booklet, *The Book of the Brownies*, begins with "The Brownie's Story of the Brownie," a fairytale about the camera's "magical" origins. Here is the relevant excerpt: "As all good children know, mortals must heed the behest of the Fairy Queen, and so the magical box was reproduced again and again and scattered over the wide world wherever Brownie bands were found, to whisper in waiting ears the secret of the pleasures in the little black box" (quoted in Olivier, "George Eastman's Modern Stone-Age Family," 10).

14. Cushing, "The Magic Black Box," 23.

15. Barnouw, *Tube of Plenty*, 8.

16. Ibid., 9.

17. Grossi, "Aviation Recorder Overview," 41.

18. "New Device Gives Check-up on Fliers," 15. See also "Flight Recorders for Commercial Transport Planes." In contemporary usage, the term *flight recorder*, like the term

black box, can designate either the flight-data or the cockpit voice recorder. In this and the next two sections, *flight recorder* refers only to the former.

19. The historical record contains some ambiguity as to the function of the latter pen. The *New York Times* article makes no mention of an automatic-pilot pen. Instead, it claims that "the third [pen] will set down the record of the radio beacon receiver whether it functions or not and for how long" ("New Device Gives Check-up on Fliers," 15). The *Science* article, on the other hand, makes no mention of a radio-beacon-receiver pen.

20. "Flight Recorders for Commercial Transport Planes," 7.

21. Ibid.

22. "New Device Gives Check-up on Fliers," 15.

23. "Flight Recorders for Commercial Transport Planes," 7.

24. "Flight Recorder to Ease Test Pilot's Job," 149. See also "Flight Recorder Takes Over Some Flight Engineer's Duties."

25. "Flight Recorder to Ease Test Pilot's Job," 149.

26. See "Wing Talk"; Giffen, "A Flight Recorder for Aircraft"; Lynn C. Thomas, "Flight Test Recorder"; and Thomson and North, "Electronic Flight Recorder."

27. Giffen, "A Flight Recorder for Aircraft," 15. *Flying* put it this way in an article published one month later: "The advantages of using radio transmission is [sic] obvious, for in testing new type airplanes it is not unusual to lose a ship in a crash and too often the cause of the failure cannot be determined. With flight test data transmitted to the ground by radio, a permanent record of what transpired is available and in all probability would prevent a duplication of the accident" (Lynn C. Thomas, "Flight Test Recorder," 61).

28. Thomson and North, "Electronic Flight Recorder," 82.

29. Burkhardt, *The Federal Aviation Administration*, 16.

30. Grossi, "Aviation Recorder Overview," 32. See also Sendzimir, "Black Box," 28.

31. "Board Is Cautious on Flight Recorders," 53.

32. Ibid., 53.

33. "Flight Recorder," *Science Digest*, 95. See also "Device Aids Plane Safety."

34. Sendzimir, "Black Box," 28.

35. "Flight Recorder," *Science Digest*, 96.

36. By the 1940s, Minneapolis-based General Mills had, in fact, become much more than a cereal company. During the Second World War, the company's Mechanical division developed and manufactured gunsights, bombsights, torpedoes, and other precision-control instruments for the U.S. Army, Navy, and Air Force. Shortly after the war, it worked on a number of government-sponsored hot-air-balloon projects, some of which concerned the atmospheric effects and implications of atomic radiation. See Forsythe, *General Mills*.

37. Shamburger, "Flight Recorder," 42. See also "Flight Data Recorder Uses No Electronics"; "Flight Recorder," *Time*; "Instrument Records Flight Performance"; and Sendzimir, "Black Box," 28–29.

38. "Instrument Records Flight Performance," 64.

39. Ryan, "The General Mills Ryan Flight Recorder," 1.

40. Ibid., 3–4.

41. Ibid., 4.

42. Ibid., 5.

43. Ibid., 8.

44. Ibid. In his paper's introduction, Ryan stated, "The cause of aircraft crashes and operational failures may be determined from an analysis of the history of the flight immediately preceding and following the difficulty. From this knowledge, corrective measures to prevent reoccurrence may be instituted. Such statistical information is necessary to safely fly air vehicle[s], and to predict the development needs of the future; thus, since only a small per cent encounter difficulty, all aircraft should carry recorders" (ibid., 1).

45. Barthes, *Mythologies*, 150.

46. I discuss the gospel of learning through failure in chapter 1.

47. da Vinci, *The Notebooks of Leonardo da Vinci*, 294–95. I am grateful to Tyler Curtain for pointing me to Leonardo.

48. Surely no military vehicle has, over the course of its history, been more emblematic of machinic indestructibility than the tank. For a cultural history of the tank, see Wright, *Tank*.

49. Sterne, *The Audible Past*, 288. See also Marvin, *When Old Technologies Were New*, 203–5.

50. Edison, "The Phonograph and Its Future," 529.

51. Sterne, *The Audible Past*, 299.

52. On the compact disc's mythos of indestructibility, see Knopper, "Are CDs Rotting Away?"

53. "If we consider sound recording on the basis of its technical possibilities, repeatability is as much a central characteristic of the technology as preservation is. In fact, the former is a prior condition of the latter" (Sterne, *The Audible Past*, 288).

54. Sendzimir, "Black Box," 29. Lockheed also produced a rectangular version, Model 109-D, shaped so as to fit into the cockpit's radio rack.

55. Significantly, the survivability thresholds specified in TSO-C51—100-g impact shock, 1100° C fire for thirty minutes, immersion in seawater for thirty-six hours—exactly matched the tolerances and endurances Ryan claimed for his invention. Its engineering and design details, its subjection to scientifically controlled destructive testing, its role as regulatory standard-bearer—all support the proposition that the General Mills Ryan Flight Recorder marked an important break, a genealogical rupture of sorts, with previous flight recorders. In this sense, it represented the first truly modern flight-data recorder.

56. Sendzimir, "Black Box," 29. See also "Flight Recorder Competition Grows as ATA Asks Deadline Extension."

57. See Grossi's "Aviation Recorder Overview" for more on the technical and regulatory history of flight recorders in the United States.

58. "Army Talk in England," 11. Though the article does not carry a byline, it was probably written by William Drysdale, whose byline appears beneath a subsequent article on the same page. Drysdale was a foreign correspondent for the *Times* during this period.

59. I am grateful to Jonathan Sterne for pointing me to this article.
60. Pingree and Gitelman, "Introduction," xii.
61. Pingree and Gitelman themselves borrow the term from Rick Altman's notion of "crisis historiography."
62. Edison, "The Phonograph and Its Future," 531–34.
63. Morton, *Off the Record*, 3.
64. Edison, "The Phonograph and Its Future," 527.
65. Ibid., 530.
66. Ibid., 530–31, 533, 536.
67. Ibid., 532, 528–29.
68. Nasaw, *Going Out*, 123.
69. Edison, "The Phonograph and Its Future," 528.
70. According to the *OED*, the transitive verb *ruggedize*—meaning "to make rugged or robust; to produce in a version designed to withstand rough usage"—first appears in the written record in 1947. *Oxford English Dictionary Online*, s.v. "ruggedize, v.," www.oed.com.
71. Menard, "Neptune's Sea-Mail Service," 337.
72. Edison, "The Phonograph and Its Future," 536.
73. Warren, *A Device for Assisting Investigation into Aircraft Accidents*, 2–3.
74. Ibid., 2.
75. Ibid., 4,
76. Ibid., 3.
77. Ibid., 4.
78. Ibid.
79. Ibid.
80. Morton, *Off the Record*, 145.
81. Morton, "Minifon," http://recording-history.org/HTML/minifon.php.
82. Witham, *Black Box*, 31.
83. Defence Science and Technology Organisation, "The Black Box," www.dsto.defence.gov.au/attachments/The%20Black%20Box.pdf.
84. "Black Box," www.abc.net.au/dimensions/dimensions_future/Transcripts/s736952.htm.
85. Defence Science and Technology Organisation, "The Black Box."
86. Job, "David Warren," 108.
87. Ibid.
88. Ibid.
89. Defence Science and Technology Organisation, "The Black Box."
90. Grossi, "Aviation Recorder Overview," 32.
91. "F.A.A. Spurs Hunt for Device to Preserve Cockpit Messages," 50.
92. Burkhardt, *The Federal Aviation Administration*, 95.
93. "Magnetic Wire Recorder," 49.
94. Ibid.
95. MacPherson, *The Black Box: All-New Cockpit Voice Recorder Accounts of In-flight Accidents*, xiv.
96. Ibid., xiii.

97. Faith, *Black Box*, 18.

98. "Witness to Terror," 1.

99. Terrazzano, "The Crash of Flight 587," www.newsday.com/news/the-crash-of-flight-587-voice-recorder-found-1.765988.

100. Faith, *Black Box*, 19.

101. Raplee, "And I Alone Survived," 84.

102. According to MacPherson, "The flight recorders were recovered in the deep on 28 February 1996, at a cost of $1.4 million" (*The Black Box: All-New Cockpit Voice Recorder Accounts of In-flight Accidents*, 49).

103. "Witness to Terror," 6.

104. The blurb on the back cover of MacPherson's *The Black Box: All-New Cockpit Voice Recorder Accounts of In-flight Accidents* states, "Anyone who watches the news knows about the 'black box.'"

105. Rimer, "For Kin of Some Jet Crash Victims, Not Hearing Last Tape Is Too Much to Bear," www.nytimes.com/1996/12/01/us/for-kin-of-some-jet-crash-victims-not-hearing-last-tape-is-too-much-to-bear.html.

106. Penley, NASA/TREK, 59–60.

107. Picker, *Victorian Soundscapes*, 114.

108. Quoted in ibid., 116.

109. Serres, *The Parasite*, 6.

110. Ibid., 12.

111. Witham, *Black Box*, 35.

112. Kittler, *Gramophone, Film, Typewriter*, 70.

113. Witham, *Black Box*, 35.

114. Ibid., 36.

115. I am referring to the situation in the United States, but the procedures are similar in other countries with modern aviation industries and bureaucracies.

116. Leave it to *Newsweek* to lay it on thick: "The terrorists had years to plan their hijacking. The passengers had just minutes to respond. But a band of patriots came together to defy death and save a symbol of freedom" (Breslau, Clift, and Thomas, "The Real Story of Flight 93," 54).

And this: "America's latest war for freedom did not begin with a speech by George W. Bush or a cruise-missile attack on a terrorist-training camp in Afghanistan. It began with a group of citizen soldiers on Flight 93 who rose up, like their forefathers, to defy tyranny. And when they came storming down the aisle, it wasn't the Americans who were afraid. It was the terrorists" (ibid., 56).

In addition to interviewing victims' family members and friends, *Newsweek* claims to have been in touch with "informed sources [who] described in detail . . . the words and sounds picked up by the cockpit voice recorder on Flight 93, information that has never been revealed before. The tapes shed light on a central mystery: they strongly suggest that the hijackers flew the plane into the ground under ferocious assault from the passengers. The picture is one of shock, struggle, fear—but the lasting impression that remains is of courage, the kind of extraordinary bravery ordinary Americans can show" (ibid.).

117. MacPherson, *The Black Box: All-New Cockpit Voice Recorder Accounts of In-flight Accidents*, xiv.

118. Quoted in Faith, *Black Box*, 27.

119. Sendzimir, "Black Box," 34.

120. Lake, "Airplane Flight Recorders," www.nytimes.com/2000/03/09/technology/airplane-flight-recorders-the-boxes-that-live-to-tell-the-tale.html.

121. Serres, *The Parasite*, 12.

122. Sterne, *The Audible Past*, 23.

123. Kittler, *Discourse Networks 1800/1900*, 236.

124. Kittler, *Gramophone, Film, Typewriter*, 14.

125. Ibid., 15–16.

126. Larkin, *Signal and Noise*, 239.

127. Serres, *The Parasite*, 18.

128. Ronell, *The Telephone Book*, 350.

129. Derrida, *Of Grammatology*. Gayatri Chakravorty Spivak, in her remarkable preface to *Of Grammatology*, provides a rich explication of the term.

130. Edison, "The Phonograph and Its Future," 533–34.

131. Johnson, "A Wonderful Invention," 304.

132. Peters, *Speaking into the Air*, 162.

133. Guthke, *Last Words*, 6.

134. Peters, *Speaking into the Air*, 223.

135. Plato, *The Collected Dialogues, Including the Letters*, 24.

136. See Kastenbaum, "Deathbed Scenes."

137. Guthke, *Last Words*, 29.

138. Quick, "Some Reflections on Dying Declarations," 109; Koerner, "Last Words," www.legalaffairs.org/issues/November-December-2002/review_koerner_novdec2002.msp.

139. Kastenbaum, "Last Words," 273.

140. The authenticity, integrity, and legitimacy of these audio files are open to question: what is their provenance? Have they been edited or otherwise manipulated? Are they tantamount to bootlegs?

141. U.S. Code, 2006 Edition, Supplement 3, Title 49—Transportation, Section 1114—Disclosure, availability, and use of information.

142. Levin, "Flight 93 Families Hear Cockpit Tape," http://old.post-gazette.com/nation/20020419flight930419p2.asp.

143. Ibid.

144. Ibid.

145. U.S Code, Supplement 3, Title 49, Section 1114.

146. Sturkey, *Mayday*, 4–5.

147. Unless otherwise noted, all quotes in this paragraph are from one of the following: MacPherson, *The Black Box: Cockpit Voice Recorder Accounts of Nineteen In-Flight Accidents*; MacPherson, *The Black Box: All-New Cockpit Voice Recorder Accounts of In-flight Accidents*; Sturkey, *Mayday*; Sturkey, *Mid-Air*.

148. MacPherson, interview by Ira Glass, www.thisamericanlife.org/radio-archives/episode/114/transcript.

chapter four. TESTS AND SPLIT SECONDS

For an extended treatment of the test drive, see Ronell, *The Test Drive*.

1. Severy, "Automobile Collisions on Purpose," 186.
2. Schivelbusch, *The Railway Journey*, 14.
3. Quoted in ibid., 14.
4. Quoted in Eastman, *Styling vs. Safety*, 115.
5. Quoted in ibid.
6. Quoted in Jain, "'Dangerous Instrumentality,'" 87 n. 16.
7. Quoted in Flink, *The Automobile Age*, 138–39.
8. Jain, "'Dangerous Instrumentality,'" 71.
9. Eastman, *Styling vs. Safety*, 116.
10. Ibid., 118.
11. Quoted in Jain, "'Dangerous Instrumentality,'" 87 n. 16.
12. Ibid., 70. "The outcome of these cases ultimately determined that automobiles would be considered as fundamentally benign products that were harmful only when driven negligently, and by the second decade of the 20th century, safety bureaucrats concentrated their attention exclusively on driver education and traffic engineering" (ibid., 65).
13. Ibid., 70.
14. Ewald, "Insurance and Risk," 202.
15. Jain notes that in the early twentieth century, injuries inflicted on bystanders "were used by automobile associations, courts, and an emerging safety network to make moral claims on drivers" ("'Dangerous Instrumentality,'" 66).
16. By "government regulation," I mean regulation of the automobile industry. The industry was actually in favor of government regulation of drivers and roads. This position was consistent with how it framed the problem of the accident and with what it took to be its economic self-interest.
17. Eastman, *Styling vs. Safety*, 119.
18. Ibid., 120–21.
19. Ibid., 122.
20. Ibid.
21. Ibid.
22. Ibid., 124.
23. Furnas, "—And Sudden Death," 21.
24. Ibid., 23.
25. Ibid., 22–23.
26. Ibid., 22.
27. Beckman, *Crash*, 115.
28. Eastman, *Styling vs. Safety*, 138.
29. Furnas, "—And Sudden Death," 21.
30. Ibid.
31. Ibid.
32. Eastman, *Styling vs. Safety*, 137.
33. Furnas, "—And Sudden Death," 26.

34. Ibid., 22.
35. Ibid., 23.
36. For present purposes, I am leaving aside the question of roadway conditions. The idea that accidents were caused by bad roads was not invoked nearly as often or as insistently as the idea that they were caused by bad drivers.
37. Hayles, *How We Became Posthuman*, 85–86.
38. Ibid., 86.
39. Jain, "'Dangerous Instrumentality,'" 70.
40. Ibid., 71.
41. Feenberg, *Critical Theory of Technology*, 5.
42. Ibid.
43. Winner, *The Whale and the Reactor*, 5–6.
44. Ibid., 6.
45. Feenberg, *Critical Theory of Technology*, 5.
46. Nader, *Unsafe at Any Speed*.
47. Jain, "'Dangerous Instrumentality,'" 66.
48. Eastman, *Styling vs. Safety*, 181–84.
49. Quoted in ibid., 181.
50. Straith actually patented three types of automobile crash padding between 1935 and 1937.
51. Eastman, *Styling vs. Safety*, 182.
52. Ibid., 184.
53. Woodward, "Medical Criticism of Modern Automotive Engineering," 628–29.
54. Woodward, like Straith, called for a car equipped with, among other things, "a latch to lock the backs of front seats in position . . . to prevent the added impact of them to the front seat passengers in sudden stops," seatbelts, "a hydraulic steering column which will move forward under a force of approximately 100 foot pounds," sponge-rubber crash padding "on the dash and back of front seats," and non-projecting knobs, buttons, and handles (ibid., 630).
55. Beckman, *Crash*, 125.
56. De Haven, "Mechanical Analysis of Survival in Falls from Heights of Fifty to One Hundred and Fifty Feet," 586 fn.
57. Hasbrook, "The Historical Development of the Crash-Impact Engineering Point of View," 269.
58. Eastman, *Styling vs. Safety*, 211.
59. De Haven, "Mechanical Analysis of Survival in Falls from Heights of Fifty to One Hundred and Fifty Feet," 586 fn.
60. Nader, *Unsafe at Any Speed*, 82.
61. De Haven defines a "free fall" as "a fall free of any obstruction other than that encountered at its termination" ("Mechanical Analysis of Survival in Falls from Heights of Fifty to One Hundred and Fifty Feet," 588).
62. Ibid., 586.
63. Ibid., 596.
64. Ibid., 587.

65. Beckman, *Crash*, 126. See also Eastman, *Styling vs. Safety*, 214. The project was headquartered in the Department of Physiology.

66. Hasbrook, "The Historical Development of the Crash-Impact Engineering Point of View," 271.

67. De Haven, "Cushion That Impact!," 10.

68. De Haven wrote, "Until the Crash Injury Research project was initiated the philosophy behind accident investigation was solely to find causes of accidents; this philosophy fell short of finding causes of injury and fatality in survivable accidents and left many problems of safety untouched. No organized attack could be made on basic problems of reducing the overall seriousness of injury in accidents without knowing what caused typical crash injuries" ("Crash Injury Research," v).

69. Ibid., v.

70. Eastman, *Styling vs. Safety*, 222.

71. De Haven, "Crash Injury Research," v.

72. Ibid.

73. One G is equivalent to the force of the earth's gravity at sea level.

74. "The Fastest Man on Earth," 85.

75. Ibid., 86.

76. NASA History Office, "History of Research in Space Biology and Biodynamics at the U.S. Air Force Missile Development Center, Holloman Air Force Base, New Mexico, 1946–1958," www.hq.nasa.gov/office/pao/History/afspbio/part4-1.htm.

77. According to the NASA History Office, "The high-speed escape problem was one of imposing magnitude. A pilot bailing out at transonic or supersonic speed had to face first the ejection force required to get him out of his plane, then the sudden onslaught of windblast and wind-drag deceleration, likely to be followed by dangerous tumbling and spinning. Any one of these forces taken separately was a potential cause of injury or death, not to mention the anxiety on the part of aircraft pilots who did not know if they would survive or not in case of ejection" (ibid.).

78. Ibid.

79. Stapp, "Human Factors of Crash Protection in Automobiles," 3.

80. NASA History Office, "History of Research in Space Biology and Biodynamics at the U.S. Air Force Missile Development Center, Holloman Air Force Base, New Mexico, 1946–1958," www.hq.nasa.gov/office/pao/History/afspbio/part5-7.htm.

81. Ibid.

82. "The Fastest Man on Earth," 86.

83. Stapp, "Human Factors of Crash Protection in Automobiles," 1.

84. Dye, "Automobile Crash Safety Research," 1.

85. De Haven was already experimentally "fooling around" with falling eggs and "a soft sponge rubber mat" by June 1936. De Haven quoted in Beckman, *Crash*, 126.

86. Dye, "Automobile Crash Safety Research," 2. See also Dye, "Cornell University Tests Show Just What Happens in a Crash . . . and How to Protect Yourself," 86.

87. Dye, "Automobile Crash Safety Research," 3–4. See also De Haven, "Cushion That Impact!," 11.

88. Dye, "Automobile Crash Safety Research," 7–8.

89. Unlike full-scale crash tests, these crash-snubbing tests used a two-inch-thick, 200-foot-long steel cable to abruptly stop the car—"a violent form of crack-the-whip," in Dye's words ("Cornell University Tests Show Just What Happens in a Crash . . . and How to Protect Yourself," 85).

90. Ibid., 33.

91. Ibid., 85.

92. The seatbelt kit and steering-wheel cushion each were available by mail order from a manufacturer in Rochester, New York, while the dashboard-pad materials could be acquired at the local hardware store (or so it was suggested) or through the mail from a manufacturer in New York City.

93. Dye, "Cornell University Tests Show Just What Happens in a Crash . . . and How to Protect Yourself," 33.

94. Ibid., 85–86.

95. Ibid., 86.

96. Ibid., 85.

97. Ibid., 33.

98. Both media coverage of Stapp's experiments and ITTE crash-safety films frequently interpolated slow-motion spectacles of deceleration/destructive testing into their visual discourse. Dye's article, by contrast, had to rely on serial photographs and serial illustrations of crashing cars and airborne dummies, as well as provocative verbal characterizations of those cars and dummies, to arouse interest and sustain attention. Consider, for instance, this phrase: "Those dummies move with amazing violence when they don't have the benefit of safety belts" (ibid., 85). And this one, quoted previously: "The blows sustained in so brief a time would amaze you."

99. Eastman, *Styling vs. Safety*, 191. Eastman does not mention the crash demonstration's effects on the graduate student. Let that sink in, too.

100. Severy, "Automobile Collisions on Purpose," 186.

101. Severy, "Automobile Crash Effects," 2.

102. See Roth, "Physical Factors Involved in Head-on Collisions of Automobiles."

103. Severy, "Automobile Crash Effects," 2.

104. Severy, Mathewson, and Siegel, *Statement on Crashworthiness of Automobile Seat Belts for the Subcommittee of the Committee on Interstate and Foreign Commerce of the House of Representatives*, Institute of Transportation and Traffic Engineering, University of California, Los Angeles, 1957, 1.

105. Severy, "Automobile Crash Effects," 4.

106. Ibid., 5, emphasis omitted.

107. The historical record indicates that at least one preliminary experiment was conducted in 1953, probably in the last quarter of the year. See "Automobile Safety Belt Undergoes Crash Tests," A8.

108. Respectively: Severy, Mathewson, and Siegel, *Statement on Crashworthiness of Automobile Seat Belts for the Subcommittee of the Committee on Interstate and Foreign Commerce of the House of Representatives*, 7; Niquette, "Engineered Automobile Crashes," 14.

109. Niquette, "Engineered Automobile Crashes," 14–15.

110. Gerlough, *Instrumentation for Automobile Crash Injury Research*, 3.

111. "One of 2 Dummies in Car Crash Test Loses Head," 9.

112. "Dummy 'Killed' in Smashed-up Car Could Be You," 102.

113. Ibid.

114. Bierman and Larsen, "Reactions of the Human to Impact Forces Revealed by High Speed Motion Picture Technique," 408.

115. Stapp, "Human Exposures to Linear Deceleration," 3–5.

116. Severy, "Automobile Crash Effects," 4, emphasis omitted.

117. Severy, "Photographic Instrumentation for Collision Injury Research," 69.

118. Ibid.

119. I discuss Murphy's Law in this book's introduction.

120. Severy, "Photographic Instrumentation for Collision Injury Research," 72.

121. Ibid., 69.

122. Ibid., 74–75.

123. Ibid., 75.

124. Ibid., 73.

125. Ibid., 75.

126. Ibid., 69.

127. Stonex and Skeels, "A Summary of Crash Research Techniques Developed by the General Motors Proving Ground."

128. Gerald M. Wilson (General Motors Proving Grounds Communications Manager), telephone interview by author, October 2003. See also "Crash Testing," *Modern Marvels*, The History Channel, 1999.

129. Stonex and Skeels, "A Summary of Crash Research Techniques Developed by the General Motors Proving Ground," 7.

130. Ibid.

131. Severy, "Photographic Instrumentation for Collision Injury Research," 69. The second and third sentences of this quote appear in an asterisked footnote.

132. On Marey, Muybridge, and the Gilbreths, see Braun, *Picturing Time*; and Doane, *The Emergence of Cinematic Time*.

133. Severy, "Photographic Instrumentation for Collision Injury Research," 75.

134. The term *second collision* was popularized, though not coined, by Ralph Nader in *Unsafe at Any Speed*.

135. Severy, "Automobile Collisions on Purpose," 186.

136. Curtis, "Still/Moving," 228.

137. On technology and "the will to mastery," see Heidegger, *Basic Writings*, 289. See also the first epigraph to this book's introduction.

138. Severy, Mathewson, and Siegel, *Auto Crash Studies*. It is noteworthy that funding for the experiments described in *Auto Crash Studies* was provided by the U.S. Air Force at the behest of John Paul Stapp.

139. Ibid., I-87.

140. Castel, "From Dangerousness to Risk," 289.

141. Respectively: quoted in Schivelbusch, *The Railway Journey*, 131; Severy, Mathewson, and Siegel, *Auto Crash Studies*, I-9; ibid., I-87–I-90.

epilogue. RETROSPECTIVE PROPHECIES

1. Huxley, *Science and the Hebrew Tradition*, 8.
2. Ibid., 13, 9.
3. *Oxford English Dictionary Online*, s.v. "palaetiology, n.," www.oed.com.
4. Huxley, *Science and the Hebrew Tradition*, 5–6
5. Ibid., 6.
6. Whewell, *History of the Inductive Sciences from the Earliest to the Present Time*, 499.
7. Ibid., 500, 499.
8. Ibid., 501.
9. Ibid., 503.
10. Huxley, *Science and the Hebrew Tradition*, 7.
11. Derrida, *Specters of Marx*, 57.
12. Limpert, *Motor Vehicle Accident Reconstruction and Cause Analysis*, 3-1.
13. Fricke and Baker, "Process of Traffic Crash Reconstruction," 2.
14. Hicks, "Traffic Accident Reconstruction," 129.
15. Limpert, *Motor Vehicle Accident Reconstruction and Cause Analysis*, 37-2.
16. Rivers, *Traffic Accident Investigators' Manual*, 3.
17. Limpert, *Motor Vehicle Accident Reconstruction and Cause Analysis*, 37-3.
18. Rivers, *Traffic Accident Investigators' Manual*, 45.
19. Van Kirk, *Vehicular Accident Investigation and Reconstruction*, 25–44.
20. Limpert, *Motor Vehicle Accident Reconstruction and Cause Analysis*, 37-19.
21. Sneddon, "Measuring at the Scene of Traffic Collisions," 216, 217.
22. Ibid., 233.
23. On the difference between mechanical modes of reproduction and manual ones, see Lynch, "Science in the Age of Mechanical Reproduction."
24. Fricke and Baker, "Photographing the Collision Scene and Damaged Vehicles," 258, emphasis omitted.
25. Ibid., 262.
26. Benjamin, *The Work of Art in the Age of Its Technological Reproducibility*, 278, 279, 277. Rosalind Krauss has rightly noted the limitations and imprecision of Benjamin's analogy. See her book *The Optical Unconscious*, 178–79.
27. Sontag, *On Photography*, 120–21.
28. Ibid., 119.
29. Fricke and Baker, "Photographing the Collision Scene and Damaged Vehicles," 258, emphasis omitted. Michael Lynch puts it this way: "The photograph preserves moments in a past that would otherwise be forgotten, and provides an examinable record of details that would otherwise go unseen and unnoticed" ("Science in the Age of Mechanical Reproduction," 214).
30. As Bruno Latour says, "A new visual culture redefines both what it is to see, and what there is to see" ("Drawing Things Together," 30).
31. Van Kirk, *Vehicular Accident Investigation and Reconstruction*, 72, emphasis added.
32. Kawin, *Mindscreen*, xi; Freud, *The Interpretation of Dreams*, 538.
33. Gaboriau, "The Little Old Man of Batignolles," 49.
34. See Vertov, *Kino-Eye*.

35. Fricke and Baker, "Photographing the Collision Scene and Damaged Vehicles," 258.
36. Ibid., 277.
37. Ibid., 259.
38. Ibid., 259–60.
39. Lynch, "Science in the Age of Mechanical Reproduction," 214, emphasis omitted.
40. Ray, *How a Film Theory Got Lost and Other Mysteries in Cultural Studies*, 20.
41. Sontag writes, "Surrealism has always courted accidents, welcomed the uninvited, flattered disorderly presences. What could be more surreal than an object which virtually produces itself, and with a minimum of effort? An object whose beauty, fantastic disclosures, emotional weight are likely to be further enhanced by any accidents that might befall it?" (*On Photography*, 52–53).
42. Benjamin, *The Work of Art in the Age of Its Technological Reproducibility*, 276.
43. On Benjamin's interest in detective fiction, see Salzani, "The City as Crime Scene."
44. Barthes, *Image, Music, Text*, 52–68.
45. Ibid., 53, 55
46. Ibid., 54, 65; Stephen Heath, "Translator's Note," 9–11.
47. Roland Barthes, *Camera Lucida*, passim.
48. Ibid., 27, 26.
49. Ibid., 42, 27.
50. Ray, *How a Film Theory Got Lost and Other Mysteries in Cultural Studies*, 20.
51. Ibid., 21.
52. Sontag, *On Photography*, 117.
53. Ray, *How a Film Theory Got Lost and Other Mysteries in Cultural Studies*, 20; Barthes, *Camera Lucida*, 33.
54. Ibid., 85.
55. Ginzburg, *Clues, Myths, and the Historical Method*; Rabaté, *Given*. Messac is the author of *The "Detective Novel" and the Influence of Scientific Thought* (1929), which, so far as I know, has never been translated into English.
56. Remer, ed., *Serendipity and the Three Princes*.
57. Ginzburg, *Clues, Myths, and the Historical Method*, 102–3.
58. Ibid., 116.
59. Quoted in ibid.

bibliography

Adams, Charles Francis, Jr. *Notes on Railroad Accidents*. New York: G. P. Putnam's Sons, 1879.
Adams, Henry. *The Education of Henry Adams*. New York: Modern Library, 1931.
Adas, Michael. *Machines as the Measure of Men: Science, Technology, and Ideologies of Western Dominance*. Ithaca: Cornell University Press, 1989.
Adorno, Theodor W. "Progress." *Benjamin: Philosophy, Aesthetics, History*, ed. Gary Smith, 84–101. Chicago: University of Chicago Press, 1989.
Aldrich, Mark. *Death Rode the Rails: American Railroad Accidents and Safety, 1828–1965*. Baltimore: Johns Hopkins University Press, 2006.
Aristotle. *Aristotle's Physics: A Revised Text with Introduction and Commentary*. Translated by W. D. Ross. London: Oxford University Press, 1955.
———. *Metaphysica*. Translated by W. D. Ross. Oxford: Clarendon, 1908.
"Army Talk in England." *New York Times*, 16 September 1888, 11.
"Automobile Safety Belt Undergoes Crash Tests." *Los Angeles Times*, 30 November 1953, A8.
Babbage, Charles. *Passages from the Life of a Philosopher*. Edited by Martin Campbell-Kelly. London: Pickering and Chatto, 1991.
Barnouw, Erik. *Tube of Plenty: The Evolution of American Television*. 2d rev. edn. New York: Oxford University Press, 1990.
Barstow, G. Forrester. Preface. *Railroad Accidents: Their Causes and the Means of Preventing Them*, by Emile With, trans. G. Forrester Barstow, v–xi. Boston: Little, Brown, 1856.
Barthes, Roland. *Camera Lucida: Reflections on Photography*. Translated by Richard Howard. New York: Hill and Wang, 1981.
———. *The Grain of the Voice: Interviews 1962–1980*. Translated by Linda Coverdale. New York: Hill and Wang, 1985.
———. *Image, Music, Text*. Translated by Stephen Heath. New York: Hill and Wang, 1977.
———. *Mythologies*. Translated by Annette Lavers. New York: Noonday Press / Farrar, Straus and Giroux, 1972.
Baudrillard, Jean. *Symbolic Exchange and Death*. London: Sage, 1993.
Bauman, Zygmunt. *Modernity and Ambivalence*. Cambridge: Polity, 1991.
Beck, Ulrich. *Risk Society: Towards a New Modernity*. Translated by Mark Ritter. London: Sage, 1992.

Beckman, Karen. *Crash: Cinema and Politics of Speed and Stasis.* Durham: Duke University Press, 2010.

Benjamin, Walter. *The Work of Art in the Age of Its Technological Reproducibility, and Other Writings on Media.* Edited by Michael W. Jennings, Brigid Doherty, and Thomas Y. Levin. Translated by Edmund Jephcott, Rodney Livingstone, Howard Eiland, et al. Cambridge: Belknap Press / Harvard University Press, 2008.

Bennett, Carl. *The Black Box.* Silent Era, "Progressive Silent Film List," www.silentera.com/PSFL/data/B/BlackBox1915.html.

Bierman, Howard R., and Victor R. Larsen. "Reactions of the Human to Impact Forces Revealed by High Speed Motion Picture Technique." *Journal of Aviation Medicine* 17, no. 5 (1946): 407–12.

"Black Box." *New Dimensions.* Australian Broadcasting Corporation. 27 November 2002. www.abc.net.au/dimensions/dimensions_future/Transcripts/s736952.htm.

Bloch, Ernst. *Literary Essays.* Translated by Andrew Joron et al. Stanford: Stanford University Press, 1998.

"Board Is Cautious on Flight Recorders." *Aviation Week,* 12 July 1948, 52–53.

Brain, Robert M. "Representation on the Line: Graphic Recording Instruments and Scientific Modernism." *From Energy to Information: Representation in Science and Technology, Art, and Literature,* ed. Bruce Clarke and Linda Dalrymple Henderson, 155–77. Stanford: Stanford University Press, 2002.

Braun, Marta. *Picturing Time: The Work of Etienne-Jules Marey (1830–1904).* Chicago: University of Chicago Press, 1992.

Breslau, Karen, Eleanor Clift, and Evan Thomas. "The Real Story of Flight 93." *Newsweek,* 3 December 2001.

Brockmann, R. John. *Twisted Rails, Sunken Ships: The Rhetoric of Nineteenth Century Steamboat and Railroad Accident Investigation Reports, 1833–1879.* Amityville, NY: Baywood, 2005.

Brown, E. H. *Structural Analysis, Volume 1.* New York: John Wiley and Sons, 1967.

Brunvand, Jan Harold. *Encyclopedia of Urban Legends.* New York: W. W. Norton, 2001.

———. *The Vanishing Hitchhiker: American Urban Legends and Their Meanings.* New York: W. W. Norton, 1981.

Bureau d'Enquêtes et d'Analyses pour la Sécurité de l'Aviation Civile. *Final Report on the Accident on 1st June 2009 to the Airbus A330-203, Registered F-GZCP, Operated by Air France, Flight AF 447, Rio de Janeiro–Paris,* July 2012. Le Bourget Cedex, France.

Burkhardt, Robert. *The Federal Aviation Administration.* New York: Frederick A. Praeger, 1967.

Butler, Samuel. *Prose Observations.* Edited by Hugh de Quehen. Oxford: Clarendon, 1979.

Calinescu, Matei. *Five Faces of Modernity: Modernism, Avant-Garde, Decadence, Kitsch, Postmodernism.* Durham: Duke University Press, 1987.

Carper, Kenneth L. Foreword. *Forensic Engineering: Learning from Failures,* ed. Kenneth L. Carper, iii. New York: American Society of Civil Engineers, 1986.

———. "What Is Forensic Engineering?" *Forensic Engineering,* 2d edn., 1–13. Boca Raton: CRC Press, 2001.

Castel, Robert. "From Dangerousness to Risk." *The Foucault Effect: Studies in Govern-*

mentality, ed. Graham Burchell, Colin Gordon, and Peter Miller, 281–99. Chicago: University of Chicago Press, 1991.

Clark, Nicola. "Bodies from 2009 Air France Crash Are Found." *New York Times*, 4 April 2011. http://www.nytimes.com/2011/04/05/world/europe/05brazil.html.

———. "Data Recovered from Air France Flight Recorders." *New York Times*, 16 May 2011. www.nytimes.com/2011/05/17/world/europe/17airfrance.html.

———. "Second Black Box Found in Air France Crash." *New York Times*, 3 May 2011. www.nytimes.com/2011/05/04/world/europe/04airfrance.html.

———. "Silence Still from Resting Place of Air France Recorder." *New York Times*, 30 June 2009. http://query.nytimes.com/gst/fullpage.html?res=9B02E7D61E3DF933A05755C0A96F9C8B63.

Cooter, Roger. "The Moment of the Accident: Culture, Militarism and Modernity in Late-Victorian Britain." *Accidents in History: Injuries, Fatalities and Social Relations*, ed. Roger Cooter and Bill Luckin, 107–58. Amsterdam: Rodopi, 1997.

Cooter, Roger, and Bill Luckin. "Accidents in History: An Introduction." *Accidents in History: Injuries, Fatalities and Social Relations*, ed. Roger Cooter and Bill Luckin, 1–16. Amsterdam: Rodopi, 1997.

"Crash of Flight 447." *Nova*. PBS. Aired 16 February 2011. Transcript. www.pbs.org/wgbh/nova/space/crash-flight-447.html.

"Crash Testing." *Modern Marvels*. History Channel, 1999. Videocassette (VHS).

Curtis, Scott. "Still/Moving: Digital Imaging and Medical Hermeneutics." *Memory Bytes: History, Technology, and Digital Culture*, ed. Lauren Rabinovitz and Abraham Geil, 218–54. Durham: Duke University Press, 2004.

Cushing, Charles Phelps. "The Magic Black Box." *Mentor*, June 1928.

da Vinci, Leonardo. *The Notebooks of Leonardo da Vinci*. Edited by Irma A. Richter. Oxford: Oxford University Press, 1952.

Defence Science and Technology Organisation. "The Black Box: An Australian Contribution to Air Safety." www.dsto.defence.gov.au/attachments/The%20Black%20Box.pdf.

De Haven, Hugh. *Accident Survival—Airplane and Passenger Car*. New York: Society of Automotive Engineers, 1952.

———. "Crash Injury Research: A Study Sponsored by the Navy and Air Force and Civil Aeronautics Administration: Summary Report for the Fiscal Year, July 1, 1949 to June 30, 1950." United States Air Force, Air Materiel Command. Wright-Patterson Air Force Base, Dayton, Ohio, September 1950.

———. "Cushion That Impact!" *Public Safety*, May 1950, 8–11+.

———. "Mechanical Analysis of Survival in Falls from Heights of Fifty to One Hundred and Fifty Feet." *War Medicine* 2, no. 4 (1942): 586–96.

Derrida, Jacques. *Of Grammatology*. Translated by Gayatri Chakravorty Spivak. Baltimore: John Hopkins University Press, 1997.

———. *Specters of Marx: The State of the Debt, the Work of Mourning, and the New International*. Translated by Peggy Kamuf. New York: Routledge, 1994.

"Device Aids Plane Safety." *New York Times*, 28 January 1947, 3.

Dewey, James, and Perry Byerly. "The Early History of Seismometry (to 1900)." *Bulletin of the Seismological Society of America* 59, no. 1 (1969): 183–227.

Doane, Mary Ann. *The Emergence of Cinematic Time: Modernity, Contingency, the Archive.* Cambridge: Harvard University Press, 2002.

———. "Information, Crisis, Catastrophe." *Logics of Television: Essays in Cultural Criticism,* ed. Patricia Mellencamp, 222–40. Bloomington: Indiana University Press, 1990.

Downie, Andrew, and Matthew L. Wald. "More Bodies Recovered Near Site of Plane Crash." *New York Times,* 7 June 2009. www.nytimes.com/2009/06/08/world/americas/08plane.html.

"Dummy 'Killed' in Smashed-up Car Could Be You." *Popular Science,* July 1954, 102.

Dye, Edward R. "Automobile Crash Safety Research." Medical Aspects of Traffic Accidents Conference, 4–5 May 1955, Montreal, Canada. Buffalo: Cornell Aeronautical Laboratory, 1955.

———. "Cornell University Tests Show Just What Happens in a Crash . . . and How to Protect Yourself." *Woman's Day,* November 1954, 32–36+.

Eagleton, Terry. *The Ideology of the Aesthetic.* Oxford: Blackwell, 1990.

Eastman, Joel W. *Styling vs. Safety: The American Automobile Industry and the Development of Automotive Safety, 1900–1966.* Lanham, MD: University Press of America, 1984.

Edison, Thomas A. "The Phonograph and Its Future." *North American Review* (May–June 1878): 527–36.

Evans-Pritchard, E. E. "Lévy-Bruhl's Theory of Primitive Mentality." *Journal of the Anthropological Society of Oxford* 1, no. 2 (1970): 39–60.

———. *Witchcraft, Oracles and Magic among the Azande.* Oxford: Clarendon, 1937.

Ewald, François. "Insurance and Risk." *The Foucault Effect: Studies in Governmentality,* ed. Graham Burchell, Colin Gordon, and Peter Miller, 197–210. Chicago: University of Chicago Press, 1991.

———. "Two Infinities of Risk." *The Politics of Everyday Fear,* ed. Brian Massumi, 221–30. Minneapolis: University of Minnesota Press, 1993.

"F.A.A. Spurs Hunt for Device to Preserve Cockpit Messages." *New York Times,* 6 April 1962, 50.

Faith, Nicholas. *Black Box: The Air-Crash Detectives—Why Air Safety Is No Accident.* Osceola, WI: Motorbooks, 1997.

"The Fastest Man on Earth." *Time,* 12 September 1955, 80–88.

Feenberg, Andrew. *Critical Theory of Technology.* New York: Oxford University Press, 1991.

"Flight Data Recorder Uses No Electronics." *Aviation Week,* 14 September 1953, 84.

"Flight Recorder." *Science Digest,* July 1947, 95–96.

"Flight Recorder." *Time,* 9 May 1955, 105.

"Flight Recorder Competition Grows as ATA Asks Deadline Extension." *Aviation Week,* 6 June 1960, 38.

"Flight Recorders for Commercial Transport Planes." *Science,* 27 August 1937, supp. 7.

"Flight Recorder Takes Over Some Flight Engineer's Duties." *Scientific American,* January 1943, 41.

"Flight Recorder to Ease Test Pilot's Job." *Aviation,* November 1942, 149.

Flink, James J. *The Automobile Age.* Cambridge: Massachusetts Institute of Technology Press, 1988.

Forsythe, Tom, ed. *General Mills: Seventy-five Years of Innovation, Invention, Food, and Fun.* Minneapolis: General Mills, 2003.

Foucault, Michel. *The Birth of the Clinic: An Archaeology of Medical Perception.* Translated by A. M. Sheridan Smith. New York: Vintage, 1994.

———. *Discipline and Punish: The Birth of the Prison.* Translated by Alan Sheridan. New York: Vintage, 1995.

———. "Polemics, Politics, and Problemizations: An Interview with Michel Foucault." *The Foucault Reader,* ed. Paul Rabinow, 381–90. New York: Vintage, 2010.

———. *Politics, Philosophy, Culture: Interviews and Other Writings, 1977–1984.* Edited by Lawrence D. Kritzman. Translated by Alan Sheridan et al. New York: Routledge, 1988.

———. *Power/Knowledge: Selected Interviews and Other Writings, 1972–1977.* Edited by Colin Gordon. Translated by Colin Gordon, Leo Marshall, John Mepham, and Kate Soper. New York: Pantheon, 1980.

Freud, Sigmund. *The Interpretation of Dreams.* Translated by James Strachey. New York: Basic Books, 1955.

Fricke, Lynn B., and J. Stannard Baker. "Process of Traffic Crash Reconstruction." *Traffic Crash Reconstruction,* 2d edn., ed. Lynn B. Fricke, 1–12. Evanston, IL: Northwestern University Center for Public Safety, 2010.

Fricke, Lynn B., and Kenneth S. Baker. "Photographing the Collision Scene and Damaged Vehicles." *Traffic Collision Investigation,* 9th edn., ed. Kenneth S. Baker, 257–300. Evanston, IL: Northwestern University Center for Public Safety, 2001.

Furnas, J. C. "—And Sudden Death." *Reader's Digest,* August 1935, 21–26.

Gaboriau, Émile. "The Little Old Man of Batignolles." *Great French Detective Stories,* ed. T. J. Hale, 39–87. New York: Vanguard, 1984.

Gerlough, D. L. *Instrumentation for Automobile Crash Injury Research.* Los Angeles: Institute of Transportation and Traffic Engineering, University of California, 1954.

Giddens, Anthony. *The Consequences of Modernity.* Stanford: Stanford University Press, 1990.

———. *Modernity and Self-Identity: Self and Society in the Late Modern Age.* Stanford: Stanford University Press, 1991.

Giedion, Siegfried [sic]. *Mechanization Takes Command: A Contribution to Anonymous History.* New York: Norton, 1969.

Giffen, Harvey D. "A Flight Recorder for Aircraft." *Radio News,* April 1943, 14–17+.

Ginzburg, Carlo. *Clues, Myths, and the Historical Method.* Baltimore: John Hopkins University Press, 1989.

Gitelman, Lisa. *Always Already New: Media, History, and Data of Culture.* Cambridge: Massachusetts Institute of Technology Press, 2006.

———. *Scripts, Grooves, and Writing Machines: Representing Technology in the Edison Era.* Stanford: Stanford University Press, 1999.

Godfrey, Edward. *Engineering Failures and Their Lessons.* N.p.: n.p., 1924.

Goldhurst, William. "A Literary Source for O'Neill's 'In the Zone.'" *American Literature* 35, no. 4 (January 1964): 530–34.

Goodman, Nan. *Shifting the Blame: Literature, Law, and the Theory of Accidents in Nineteenth-Century America*. Princeton: Princeton University Press, 1998.

Gordon, J. E. *Structures: or Why Things Don't Fall Down*. New York: Da Capo, 2003.

Green, Judith. "Accidents: The Remnants of a Modern Classificatory System." *Accidents in History: Injuries, Fatalities and Social Relations*, ed. Roger Cooter and Bill Luckin, 35–58. Amsterdam: Rodopi, 1997.

———. *Risk and Misfortune: The Social Construction of Accidents*. London: Routledge, 1997.

Green, Nicholas St. John. *Essays and Notes on the Law of Tort and Crime*. Menasha: George Banta Publishing, 1933.

Greenblatt, Stephen. *The Swerve: How the World Became Modern*. New York: Norton, 2011.

Gross, Hans. *Criminal Investigation: A Practical Handbook for Magistrates, Police Officers, and Lawyers*. Translated by John Adam and J. Collyer Adam. London: Specialist Press, 1907.

Grossberg, Lawrence. *Bringing It All Back Home: Essays on Cultural Studies*. Durham: Duke University Press, 1997.

Grossi, Dennis R. "Aviation Recorder Overview." *Journal of Accident Investigation* 2, no. 1 (spring 2006): 31–42.

Guthke, Karl S. *Last Words: Variations on a Theme in Cultural History*. Princeton: Princeton University Press, 1992.

Hacking, Ian. *The Emergence of Probability: A Philosophical Study of Early Ideas about Probability, Induction and Statistical Inference*. 2d edn. Cambridge: Cambridge University Press, 2006.

———. "How Should We Do the History of Statistics?" *The Foucault Effect: Studies in Governmentality*, ed. Graham Burchell, Colin Gordon, and Peter Miller, 181–95. Chicago: University of Chicago Press, 1991.

———. *The Taming of Chance*. Cambridge: Cambridge University Press, 1990.

Hall, Stanley. *Railway Detectives: The 150-year Saga of the Railway Inspectorate*. London: Ian Allan, 1990.

Hall, Stuart. "Cultural Studies and Its Theoretical Legacies." *Cultural Studies*, ed. Lawrence Grossberg, Cary Nelson, and Paula Treichler, 277–94. New York: Routledge, 1992.

Hamilton, Ross. *Accident: A Philosophical and Literary History*. Chicago: University of Chicago Press, 2007.

Hankins, Thomas L., and Robert J. Silverman. *Instruments and the Imagination*. Princeton: Princeton University Press, 1995.

Hasbrook, A. Howard. "The Historical Development of the Crash-Impact Engineering Point of View." *Clinical Orthopaedics* 8 (1956): 268–74.

Hayles, N. Katherine. *How We Became Posthuman: Virtual Bodies in Cybernetics, Literature, and Informatics*. Chicago: University of Chicago Press, 1999.

Heath, Stephen. "Translator's Note." *Image, Music, Text*, by Roland Barthes, translated by Stephen Heath, 7–11. New York: Hill and Wang, 1977.

Heidegger, Martin. *Basic Writings*. Edited by David Farrell Krell. San Francisco: Harper and Row, 1977.

Hicks, Joel T. "Traffic Accident Reconstruction." *Forensic Engineering*, 2d edn., ed. Kenneth L. Carper, 129–58. Boca Raton: CRC Press, 2001.

Hobbes, Thomas. *Leviathan: With Selected Variants from the Latin Edition of 1668*. Edited by Edwin Curley. Indianapolis: Hackett, 1994.

Horwitz, Morton J. *The Transformation of American Law, 1780–1860*. Cambridge: Harvard University Press, 1977.

Hume, David. *The Natural History of Religion*. Edited by J. C. A. Gaskin. Oxford: Oxford University Press, 1998.

Hunter, Louis. *Steamboats on the Western Rivers: An Economic and Technological History*. Cambridge: Harvard University Press, 1949.

Huxley, Thomas H. *Science and the Hebrew Tradition: Essays*. New York: D. Appleton, 1897.

Hylton, Wil S. "The Deepest End." *New York Times Magazine*, 8 May 2011, 38–45+.

Hyman, Anthony. *Charles Babbage: Pioneer of the Computer*. Princeton: Princeton University Press, 1982.

Ihde, Don. *Technology and the Lifeworld: From Garden to Earth*. Bloomington: Indiana University Press, 1990.

"Instrument Records Flight Performance." *National Safety News*, June 1955, 64.

Jain, Sarah S. Lochlann. "'Dangerous Instrumentality': The Bystander as Subject in Automobility." *Cultural Anthropology* 19, no. 1 (2004): 61–94.

Job, Macarthur. "David Warren." *Time International (South Pacific Edition)*, 25 October 1999, 108.

Johnson, Edward H. "A Wonderful Invention: Speech Capable of Indefinite Repetition from Automatic Records." *Scientific American*, 17 November 1877, 304.

Jünger, Ernst. "On Danger." *The Weimar Republic Sourcebook*, ed. Anton Kaes, Martin Jay, and Edward Dimendberg, 369–72. Berkeley: University of California Press, 1994.

Kasson, John F. *Civilizing the Machine: Technology and Republican Values in America, 1776–1900*. New York: Hill and Wang, 1999.

Kastenbaum, Robert. "Deathbed Scenes." *Encyclopedia of Death*, ed. Robert Kastenbaum and Beatrice Kastenbaum, 97–101. Phoenix: Oryx, 1989.

———. "Last Words." *Monist* 76, no. 2 (April 1993): 270–90.

Kawin, Bruce F. *Mindscreen: Bergman, Godard, and First-Person Film*. Princeton: Princeton University Press, 1978.

Keenan, Thomas, and Eyal Weizman. "Mengele's Skull." *Cabinet* 43 (fall 2011): 61–67.

———. *Mengele's Skull: The Advent of a Forensic Aesthetics*. Berlin: Sternberg, 2012.

Kittler, Friedrich A. *Discourse Networks 1800/1900*. Translated by Michael Metteer with Chris Cullens. Stanford: Stanford University Press, 1990.

———. *Gramophone, Film, Typewriter*. Translated by Geoffrey Winthrop-Young and Michael Wutz. Stanford: Stanford University Press, 1999.

Knopper, Steve. "Are CDs Rotting Away?: 'Indestructible' Technology Shows Its Age." *Rolling Stone*, 25 June 2004. www.rollingstone.com/music/news/are-cds-rotting-away-20040625.

Koerner, Brendan I. "Last Words." *Legal Affairs*, November–December 2002. www.legalaffairs.org/issues/November-December-2002/review_koerner_novdec2002.msp.

Koolhaas, Rem. "'Life in the Metropolis' or 'The Culture of Congestion.'" *Architecture Theory since 1968*, ed. K. Michael Hays, 324–25. Cambridge: Massachusetts Institute of Technology Press, 2000. Originally published in *Architectural Design* 47, no. 5 (August 1977).

Krauss, Rosalind E. *The Optical Unconscious*. Cambridge: Massachusetts Institute of Technology Press, 1993.

———. *The Originality of the Avant-Garde and Other Modernist Myths*. Cambridge: Massachusetts Institute of Technology Press, 1985.

Lake, Matt. "Airplane Flight Recorders: The Boxes that Live to Tell the Tale." *New York Times*, 9 March 2000. www.nytimes.com/2000/03/09/technology/airplane-flight-recorders-the-boxes-that-live-to-tell-the-tale.html.

Larabee, Ann. *Decade of Disaster*. Urbana: University of Illinois Press, 2000.

Larkin, Brian. *Signal and Noise: Media, Infrastructure, and Urban Culture in Nigeria*. Durham: Duke University Press, 2008.

Latour, Bruno. "Drawing Things Together." *Representation in Scientific Practice*, ed. Michael Lynch and Steve Woolgar, 19–68. Cambridge: Massachusetts Institute of Technology Press, 1990.

———. *Pandora's Hope: Essays on the Reality of Science Studies*. Cambridge: Harvard University Press, 1999.

———. *Science in Action: How to Follow Scientists and Engineers through Society*. Cambridge: Harvard University Press, 1987.

Levin, Steve. "Flight 93 Families Hear Cockpit Tape." *Pittsburgh Post-Gazette*, 19 April 2002. http://old.post-gazette.com/nation/20020419flight930419p2.asp.

Lévi-Strauss, Claude. *Structural Anthropology*. Translated by Claire Jacobson and Brooke Grundfest Schoepf. New York: Basic Books, 1963.

Lévy-Bruhl, Lucien. *Primitive Mentality*. Translated by Lilian A. Clare. London: George Allen and Unwin, 1923.

Limpert, Rudolf. *Motor Vehicle Accident Reconstruction and Cause Analysis*. 6th edn. LexisNexis, 2008.

Lossier, Henry. *The Pathology and Therapeutics of Reinforced Concrete*. Translated by D. A. Sinclair. Ottawa: National Research Council of Canada, 1962.

Lynch, Michael. "Science in the Age of Mechanical Reproduction: Moral and Epistemic Relations between Diagrams and Photographs." *Biology and Philosophy* 6 (1991): 205–26.

MacPherson, Malcolm, ed. *The Black Box: All-New Cockpit Voice Recorder Accounts of In-flight Accidents*. New York: Quill / William Morrow, 1998.

———. *The Black Box: Cockpit Voice Recorder Accounts of Nineteen In-Flight Accidents*. New York: Quill, 1984.

———. Interview by Ira Glass. "Episode 114: Last Words." *This American Life*. WBEZ Chicago Public Media, aired 23 October 1998. Transcript. www.thisamericanlife.org/radio-archives/episode/114/transcript.

"Magnetic Wire Recorder." *Life*, 1 November 1943, 49–50.

Manby, Charles, ed. *Minutes of Proceedings of the Institution of Civil Engineers with Abstracts of the Discussions Vol. XV, Session 1855–56*. London: Institution of Civil Engineers, 1856.

Marquard, Odo. *In Defense of the Accidental: Philosophical Studies*. New York: Oxford University Press, 1991.

Marvin, Carolyn. *When Old Technologies Were New: Thinking about Electric Communication in the Late Nineteenth Century*. New York: Oxford University Press, 1988.

Marx, Leo. "The Domination of Nature and the Redefinition of Progress." *Progress: Fact or Illusion?*, ed. Leo Marx and Bruce Mazlish, 201–18. Ann Arbor: University of Michigan Press, 1996.

———. *The Machine in the Garden: Technology and the Pastoral Ideal in America*. New York: Oxford University Press, 2000.

Mazlish, Bruce, and Leo Marx. "Introduction." *Progress: Fact or Illusion?*, ed. Leo Marx and Bruce Mazlish, 1–7. Ann Arbor: University of Michigan Press, 1996.

Menard, Wilmon. "Neptune's Sea-Mail Service." *Sea Frontiers: Bulletin of the International Oceanographic Foundation* (November–December 1980): 336–40.

Meyer, Susan Sauvé. "Aristotle, Teleology, and Reduction." *Philosophical Review* 101, no. 4 (October 1992): 791–825.

Mitcham, Carl, and Eric Schatzberg, "Defining Technology and the Engineering Sciences." *Philosophy of Technology and Engineering Sciences*, ed. Anthonie Meijers, 27–63. Amsterdam: Elsevier, 2009.

Morton, David. "Minifon: An Early Portable Dictating Machine." http://recording-history.org/HTML/minifon.php.

———. *Off the Record: The Technology and Culture of Sound Recording in America*. New Brunswick: Rutgers University Press, 2000.

Mumford, Lewis. *Technics and Civilization*. Chicago: University of Chicago Press, 2010.

Nader, Ralph. *Unsafe at Any Speed: The Designed-in Dangers of the American Automobile*. New York: Grossman, 1965.

NASA History Office, "History of Research in Space Biology and Biodynamics at the U.S. Air Force Missile Development Center, Holloman Air Force Base, New Mexico, 1946–1958." http://history.nasa.gov/afspbio/top.htm.

Nasaw, David. *Going Out: The Rise and Fall of Public Amusements*. New York: Basic Books, 1993.

"New Device Gives Check-up on Fliers." *New York Times*, 4 February 1937, 15.

Nietzsche, Friedrich. *On the Genealogy of Morality*. Edited by Keith Ansell-Pearson. Translated by Carol Diethe. Cambridge: Cambridge University Press, 1997.

———. *The Portable Nietzsche*. Translated by Walter Kaufmann. New York: Viking, 1954.

Niquette, Paul. "Engineered Automobile Crashes." *California Engineer* 32, no. 7 (1954): 14–15, 32.

Nisbet, Robert. *History of the Idea of Progress*. New Brunswick: Transaction, 1994.

Noon, Randall K. *Forensic Engineering Investigation*. Boca Raton: CRC Press, 2001. doi: http://dx.doi.org/10.1201/9781420041415.

Nye, David E. *American Technological Sublime*. Cambridge: Massachusetts Institute of Technology Press, 1994.

Olivier, Marc. "George Eastman's Modern Stone-Age Family: Snapshot Photography and the Brownie." *Technology and Culture* 48 (January 2007): 1–19.

"One of 2 Dummies in Car Crash Test Loses Head." *Los Angeles Times*, 16 February 1954, 9.

Oppenheim, E. Phillips. *The Black Box*. New York: Grosset and Dunlap, 1915.

Packer, Jeremy. "Disciplining Mobility: Governing and Safety." *Foucault, Cultural Studies, and Governmentality*, ed. Jack Z. Bratich, Jeremy Packer, and Cameron McCarthy, 135–61. Albany: State University of New York Press, 2003.

Paz, Octavio. *Conjunctions and Disjunctions*. Translated by Helen R. Lane. New York: Viking, 1974.

Penley, Constance. NASA/TREK: *Popular Science and Sex in America*. London: Verso, 1997.

Peters, John Durham. "Helmholtz, Edison, and Sound History." *Memory Bytes: History, Technology, and Digital Culture*, ed. Lauren Rabinovitz and Abraham Geil, 177–98. Durham: Duke University Press, 2004.

———. *Speaking into the Air: A History of the Idea of Communication*. Chicago: University of Chicago Press, 1999.

Petroski, Henry. *Design Paradigms: Case Histories of Error and Judgment in Engineering*. Cambridge: Cambridge University Press, 1994.

———. "The Success of Failure." *Technology and Culture* 42 (April 2001): 321–28.

———. *Success through Failure: The Paradox of Design*. Princeton: Princeton University Press, 2006.

———. *To Engineer Is Human: The Role of Failure in Successful Design*. New York: Vintage, 1992.

———. *To Forgive Design: Understanding Failure*. Cambridge: Belknap Press / Harvard University Press, 2012.

Piaget, Jean. *The Child's Conception of Physical Causality*. New Brunswick: Transaction, 2001.

"The Phonograph at Sea." *Phonogram*, March–April 1893.

Picker, John M. *Victorian Soundscapes*. Oxford: Oxford University Press, 2003.

Pingree, Geoffrey B., and Lisa Gitelman. "Introduction: What's New about New Media?" *New Media, 1740–1915*, ed. Lisa Gitelman and Geoffrey B. Pingree, xi–xxii. Cambridge: Massachusetts Institute of Technology Press, 2003.

Plato. *The Collected Dialogues, Including the Letters*. Bollingen Series 71. Edited by Edith Hamilton and Huntington Cairns. Princeton: Princeton University Press, 1961.

Poe, Edgar Allan. *Complete Tales and Poems*. New York: Vintage, 1975.

Porter, Dennis. "Backward Construction and the Art of Suspense." *The Poetics of Murder: Detective Fiction and Literary Theory*, ed. Glenn W. Most and William W. Stowe, 327–40. San Diego: Harcourt Brace Jovanovich, 1983.

Quick, Charles W. "Some Reflections on Dying Declarations." *Howard Law Journal* 6 (1960): 109–34.

Rabaté, Jean-Michel. *Given: 1°Art 2° Crime: Modernity, Murder and Mass Culture*. Brighton: Sussex Academic Press, 2007.

Rabinbach, Anson. *The Human Motor: Energy, Fatigue, and the Origins of Modernity*. New York: Basic Books, 1990.

Raplee, Jack. "And I Alone Survived." *Mechanical Engineering* (March 2000): 84–86.

Ray, Robert B. *How a Film Theory Got Lost and Other Mysteries in Cultural Studies*. Bloomington: Indiana University Press, 2001.

Reed, Robert C. *Train Wrecks: A Pictorial History of Accidents on the Main Line*. New York: Bonanza, 1968.

Remer, Theodore G., ed. *Serendipity and the Three Princes: From the Peregrinaggio of 1557*. Norman: University of Oklahoma Press, 1965.

Rimer, Sara. "For Kin of Some Jet Crash Victims, Not Hearing Last Tape Is Too Much to Bear." *New York Times*, 1 December 1996. www.nytimes.com/1996/12/01/us/for-kin-of-some-jet-crash-victims-not-hearing-last-tape-is-too-much-to-bear.html.

Rivers, R. W. *Traffic Accident Investigators' Manual*. 2d edn. Springfield, IL: Charles C. Thomas, 1995.

Romanyshyn, Robert D. *Technology as Symptom and Dream*. London: Routledge, 1989.

Ronell, Avital. *The Telephone Book: Technology, Schizophrenia, Electric Speech*. Lincoln: University of Nebraska Press, 1989.

———. "The Test Drive." *Deconstruction Is/in America: A New Sense of the Political*, ed. Anselm Haverkamp, 200–220. New York: New York University Press, 1995.

———. *The Test Drive*. Urbana: University of Illinois Press, 2005.

Rosenheim, Shawn James. *The Cryptographic Imagination: Secret Writing from Edgar Poe to the Internet*. Baltimore: John Hopkins University Press, 1997.

Roth, Herman P. "Physical Factors Involved in Head-on Collisions of Automobiles." *Highway Research Board: Proceedings of the Thirty-first Annual Meeting*, 349–56. Washington: National Research Council, 1952.

Ryan, J. J. "The General Mills Ryan Flight Recorder." The American Society of Mechanical Engineers Diamond Jubilee Spring Meeting, 18–21 April 1955, Baltimore, Maryland. New York: The American Society of Mechanical Engineers, 1955.

Salzani, Carlo. "The City as Crime Scene: Walter Benjamin and the Traces of the Detective." *New German Critique* 100, vol. 34, no. 1 (winter 2007): 165–87.

Sawday, Jonathan. *The Body Emblazoned: Dissection and the Human Body in Renaissance Culture*. London: Routledge, 1995.

Scarry, Elaine. *The Body in Pain: The Making and Unmaking of the World*. New York: Oxford University Press, 1985.

Schivelbusch, Wolfgang. *The Railway Journey: The Industrialization of Time and Space in the Nineteenth Century*. Berkeley: University of California Press, 1986.

Schnapp, Jeffrey T. "Crash (Speed as Engine of Individuation)." *Modernism/Modernity* 6, no. 1 (1999): 1–49.

Sekula, Allan. "The Body and the Archive." *October* 39 (winter 1986): 3–64.

Sendzimir, Vanda. "Black Box." *American Heritage of Invention and Technology* (fall 1996): 26–36.

Serres, Michel. *The Five Senses: A Philosophy of Mingled Bodies*. Translated by Margaret Sankey and Peter Cowley. London: Continuum, 2008.

———. *Hermes: Literature, Science, Philosophy*. Edited by Josue V. Harari and David F. Bell. Baltimore: Johns Hopkins University Press, 1982.

———. *The Parasite*. Translated by Lawrence R. Schehr. Baltimore: Johns Hopkins University Press, 1982.

Severy, Derwyn M. "Automobile Collisions on Purpose." *Human Factors* 2, no. 4 (1960): 186–202.

———. "Automobile Crash Effects." Paper presented at the California State Governor's Safety Conference, Sacramento, California, 7 October 1954. Los Angeles: Institute of Transportation and Traffic Engineering, University of California, 1954.

———. "Photographic Instrumentation for Collision Injury Research." *Journal of the SMPTE* 67, no. 2 (1958): 69–77.

Severy, Derwyn M., John H. Mathewson, and Arnold W. Siegel. *Auto Crash Studies*. Los Angeles: Institute of Transportation and Traffic Engineering, University of California, 1959.

———. *Statement on Crashworthiness of Automobile Seat Belts for the Subcommittee of the Committee on Interstate and Foreign Commerce of the House of Representatives*. Los Angeles: Institute of Transportation and Traffic Engineering, University of California, 1957.

Shackelford, James F. "Failure Analysis." *The Engineering Handbook*, 2d edn., ed. Richard C. Dorf, chapter 214. Boca Raton: CRC Press, 2005. doi: http://dx.doi.org/10.1201/9781420039870.

Shamburger, Page. "Flight Recorder." *Flying*, March 1954, 42.

Shapin, Steven, and Simon Schaffer. *Leviathan and the Air-Pump: Hobbes, Boyle and the Experimental Life*. Princeton: Princeton University Press, 1985.

Singer, Ben. *Melodrama and Modernity: Early Sensational Cinema and Its Contexts*. New York: Columbia University Press, 2001.

Sneddon, James P. "Measuring at the Scene of Traffic Collisions." *Traffic Collision Investigation*, 9th edn., ed. Kenneth S. Baker, 183–255. Evanston, IL: Northwestern University Center for Public Safety, 2001.

Sontag, Susan. *On Photography*. New York: Farrar, Straus and Giroux, 1977.

Spark, Nick T. "The Fastest Man on Earth." *Annals of Improbable Research* 9, no. 5 (2003): 4–24.

Stapp, John Paul. "Human Exposures to Linear Deceleration: Part 2. The Forward-facing Position and the Development of a Crash Harness." *Air Force Technical Report 5915*, December 1951.

———. "Human Factors of Crash Protection in Automobiles." Society of Automotive Engineers Summer Meeting, 3–8 June 1956, Atlantic City, N.J., and New York: Society of Automotive Engineers, 1956.

———. "Human Tolerance to Deceleration." *American Journal of Surgery* 93, no. 4 (1957): 734–40.

Sterne, Jonathan. *The Audible Past: Cultural Origins of Sound Reproduction*. Durham: Duke University Press, 2003.

Stonex, Kenneth A., and Paul C. Skeels. "A Summary of Crash Research Techniques Developed by the General Motors Proving Ground." *General Motors Engineering Journal* 10, no. 4 (1963): 7–11.

Sturkey, Marion F. *Mayday: Accident Reports and Voice Transcripts from Airline Crash Investigations*. Plum Branch, SC: Heritage Press International, 2005.

———. *Mid-Air: Accident Reports and Voice Transcripts from Military and Airline Mid-Air Collisions*. Plum Branch, SC: Heritage Press International, 2008.

Tamár, Esther. "Possible Sources for Two O'Neill One-Acts." *Eugene O'Neill Newsletter* (winter 1982). www.eoneill.com/library/newsletter/vi_3/vi-3h.htm.

Tenner, Edward. *Why Things Bite Back: Technology and the Revenge of Unintended Consequences.* New York: Alfred A. Knopf, 1996.

Terrazzano, Lauren. "The Crash of Flight 587: Voice Recorder Found." *Newsday*, 13 November 2001. www.newsday.com/news/the-crash-of-flight-587-voice-recorder-found-1.765988.

Thomas, Lynn C. "Flight Test Recorder." *Flying*, May 1943, 60–61+.

Thomas, Ronald R. *Detective Fiction and the Rise of Forensic Science.* Cambridge: Cambridge University Press, 1999.

Thomson, George H. "American Bridge Failures: Mechanical Pathology, Considered in Its Relation to Bridge Design." *Engineering* (21 September 1888): 252–53, 294.

Thomson, Thomas B., and Willard C. North. "Electronic Flight Recorder." *Radio News*, February 1945, 25–27+.

Thorwald, Jürgen. *Crime and Science: The New Frontier in Criminology.* New York: Harcourt, Brace and World, 1967.

Tichi, Cecelia. *Shifting Gears: Technology, Literature, Culture in Modernist America.* Chapel Hill: University of North Carolina Press, 1987.

Trachtenberg, Alan. Foreword. *The Railway Journey: The Industrialization of Time and Space in the Nineteenth Century*, by Wolfgang Schivelbusch, xiii–xvi. Berkeley: University of California Press, 1986.

Trovillo, Paul V. "A History of Lie Detection." *Journal of Criminal Law and Criminology (1931–1951)* 29, no. 6 (1939): 848–81.

U.S. Code, 2006 Edition, Supplement 3, Title 49—Transportation, Section 1114—Disclosure, availability, and use of information.

Valverde, Mariana. *Law and Order: Images, Meanings, Myths.* New Brunswick: Rutgers University Press, 2006.

Van Kirk, Donald J. *Vehicular Accident Investigation and Reconstruction.* Boca Raton: CRC Press, 2001.

Van Wyck, Peter C. *Signs of Danger: Waste, Trauma, and Nuclear Threat.* Minneapolis: University of Minnesota Press, 2005.

Vertov, Dziga. *Kino-Eye: The Writings of Dziga Vertov.* Edited by Annette Michelson. Translated by Kevin O'Brien. Berkeley: University of California Press, 1984.

Virilio, Paul. *Open Sky.* Translated by Julie Rose. London: Verso, 1997.

———. *The Original Accident.* Translated Julie Rose. Cambridge: Polity, 2007.

———. *Politics of the Very Worst.* Translated by Michael Cavaliere. Edited by Sylvère Lotringer. New York: Semiotext(e), 1999.

———. "The Primal Accident." *The Politics of Everyday Fear*, ed. Brian Massumi, 211–18. Minneapolis: University of Minnesota Press, 1993.

———. *Unknown Quantity.* Translated by Chris Turner. New York: Thames and Hudson, 2003.

Ward, Joseph S. "What Is a Forensic Engineer?" *Forensic Engineering: Learning from Failures*, ed. Kenneth L. Carper, 1–6. New York: American Society of Civil Engineers, 1986.

Warren, D. R. *A Device for Assisting Investigation into Aircraft Accidents.* Melbourne: Aeronautical Research Laboratories, 1954.

Whewell, William. *History of the Inductive Sciences from the Earliest to the Present Time*. Vol. 2. New York: D. Appleton, 1897.

Wiener, Norbert. *Cybernetics: Or Control and Communication in the Animal and the Machine*. 2d edn. Cambridge: Massachusetts Institute of Technology Press, 1965.

Williams, Raymond. *The Long Revolution*. New York: Columbia University Press, 1961.

"Wing Talk." *Collier's*, 26 December 1942, 8+.

Winner, Langdon. *Autonomous Technology: Technics-Out-of-Control as a Theme in Political Thought*. Cambridge: Massachusetts Institute of Technology Press, 1977.

———. *The Whale and the Reactor: A Search for Limits in an Age of High Technology*. Chicago: University of Chicago Press, 1986.

With, Emile. *Railroad Accidents: Their Causes and the Means of Preventing Them*. Translated by G. Forrester Barstow. Boston: Little, Brown, 1856.

Witham, Janice Peterson. *Black Box: David Warren and the Creation of the Cockpit Voice Recorder*. South Melbourne: Lothian, 2005.

Witmore, Michael. *Culture of Accidents: Unexpected Knowledges in Early Modern England*. Stanford: Stanford University Press, 2001.

"Witness to Terror." *The New Detectives: Case Studies in Forensic Science*. Written transcript. Head writer Steven Zorn. Written by Lynn Waltz. Directed by David Haycox. Discovery Channel, 1997.

Wittgenstein, Ludwig. *Philosophical Investigations*, 3d edn. Translated by G. E. M. Anscombe. Englewood Cliffs, NJ: Prentice Hall, 1958.

Woodward, Fletcher D. "Medical Criticism of Modern Automotive Engineering." *Journal of the American Medical Association* 138, no. 9 (1948): 627–31.

Wright, Patrick. *Tank: The Progress of a Monstrous War Machine*. London: Faber and Faber, 2000.

Žižek, Slavoj. *The Sublime Object of Ideology*. London: Verso, 1989.

Index

Aberdeen Proving Ground (U.S. Army), 174
accident discourse, 8–14; accident definitions, 11, 17, 19–22, 49–51, 54, 82; accident theory, 47–52, 54–55; crash-injury focus, 143–44, 156–62; crash-testing and, 185, 191–92, 193–94; crashworthiness focus, 164–67, 169–71; forensic mythology and, 106; human-error focus, 148–56, 158, 166, 169, 187–89; inevitability of accidents, 15, 161, 171–74, 188, 189–90, 193–94; prospective orientation and, 87–88. *See also* Aristotle; causation and causal determinism; liability
Accidents in History (Cooter and Luckin, eds.), 14
acoustics, 66–67
actuarial sciences, 14
Adams, Charles Francis, Jr., 56–57, 61
Adams, Henry, 12–13
Adas, Michael, 27
Adorno, Theodor, 29, 82
Aeromedical Field Laboratory (Holloman Air Force Base), 163–64, 165–66, 167, 188. *See also* Stapp, John Paul
Aeronautical Research Laboratories (Defence Science and Technology Organisation, Australian Department of Defence), 116, 125–26, 127. *See also* ARL Flight Memory Recorder
Air Disaster (Job), 126
Air France Flight 447 (2009 disappearance), 3–6, 23

Aldrich, Mark, 46
American Heritage of Invention and Technology (journal), 135
American Society of Civil Engineers (ASCE), 31–32
anthropometric photography. *See* mug shots
Antonioni, Michelangelo, 211–12
Apology (Plato), 139
Arendt, Hannah, 29, 218n82
Aristotle: accident definition, 49; dismissal of the accidental, 8, 9, 50–51; substance-accident binary, 48–50, 220n75
ARL Flight Memory Recorder, 117, *119*; background noise, 134; descriptions, 117–18, 120, 125–26; goals of, 116–17, 118, 128–29; "immediate cut-out," 117, 137; surveillance concerns, 120–22, 125, 126
audile technique, 135
authenticity, 71–72, 81; cockpit voice recording and, 129–30, 132, 136; indexicality and, 22–23, 81, 129, 132
auto-accident discourse. *See* accident discourse; reconstruction of auto accidents
Auto Crash Studies (ITTE), 193–94
automaticity: dangers of, 207; flight-data recorders and, 97–101, 104–5, 116, 121, 126; graphic method and, 67–68, 71–72, 105; importance of, 71–72, 81; phonautograph and, 66; photography and, 209–10
Automotive Crash Forces project (Aeromedical Field Laboratory), 166
autoptic vision, 16, 24, 37, 39, 161, 197. *See*

also forensic gaze; forensics and medical/mechanical pathology
Aviation (magazine), 98
Aviation Week (magazine), 102
Azande of central Africa, 9–10

Babbage, Charles: broad-gauge rail advocacy, 74–75, 83, 87, 223n27; data survivability, 115; general adoption of self-registering apparatus, 224n55; goals of self-registering apparatus, 66, 74; graphic-method advantages, 71; railway-safety concerns, 74–76, 79, 83. *See also* self-registering apparatus (Babbage)
Baker, Kenneth: photography's utilities, 202–3, 204–5, 206, 211; photography's infelicities, 206–8, 209
Barnouw, Erik, 94–95
Barstow, G. Forrester, 19–20
Barthes, Roland: causal determinism, 17; indexicality, 23; inoculation, 105–6; photography, 195, 210, 211
Baudrillard, Jean, 63
Bauman, Zygmunt, 11
Beck, Ulrich, 13
Beckman, Karen, 23, 151, 159, 160
Bell, Alexander Graham, 113
Benjamin, Walter, 13, 143, 204, 210
The Birth of the Clinic (Foucault), 24, 39
black boxes, 4–6, 90; cultural and historical precedents, 90–95; definitions, 89–90, 95–96, 224–25n18; early examples, 97–102; forensic mythology and, 106–7; indestructibility/permanence and, 103–4, 109–10; surveillance and, 120–22, 125. *See also* cockpit voice recorders; flight-data recorders
Black Box (Faith), 130–31
The Black Box (1915 film), 90, *91*, *93*
The Black Box (MacPherson), 129–30, 134, 136, 137, 140, 141
The Black Box (Oppenheim), 90–93
Bloch, Ernst: accident characterization, 194; detective fiction, 41, 42; engineer's anxiety, 61–63, 64

Blow-Up (1966 film), 211–12, *212*
Boethius, 8
boiler explosions, 45, 46–47, 52–53, 58–59
Boiler Explosions and How to Prevent Them (Peschka), 59
Boswell, Walter, 126
Brain, Robert, 69, 72
Brockmann, John, 55, 61, 222n126
Brown, Eric Hugh, 31
Brown Instrument Company, 98, 126
Brunvand, Jan Harold, 52
Burke, Edmund, 76–77, 78
Burkhardt, Robert, 126
Bury, J. B., 26
Butler, Samuel, 17

California Engineer (journal), 179–80
Calinescu, Matei, 7
Calvin, Jean, 8
Camera Lucida (Barthes), 195, 210, 211
Carlyle, Thomas, 12
Carper, Kenneth, 32, 38
Castel, Robert, 193–94
causation and causal determinism: accident definitions and, 11, 20–22, 49–51, 54, 82; auto-accident discourse (early), 149–50, 152, 153, 155–56; auto-accident injuries and, 160–61, 166–67, 171, 181–82; black boxes and, 5–6, 130; boiler explosions and, 47; causal instinct, 216–17n57; childlike and "primitive" conceptions of, 9–11; cockpit voice recorders and, 126, 134; early flight-data recorders and, 98, 102, 117, 118; forensic approach and, 23–24, 31–32, 43–44, 64; historical precedents to forensic engineering and, 37, 40, 41, 87–88; indexicality and, 16, 71; liability and, 53–54, 56, 59, 148, 181–82; palaetiology, 195–97; reason/scientific rationality and, 10–11, 14, 17–18, 19–21, 197; reconstruction and, 59, 97, 98; recurrence prevention and, 7, 59–60, 61, 80, 96, 105, 118, 128–29; theological approaches, 16–18
Cheysson, Ernest, 69

The Child's Conception of Physical Causality (Piaget), 9
Chrysler Corporation, 157
Civil Aeronautics Board (CAB), 101–2, 109, 160. *See also* Federal Aviation Administration (FAA)
Clinical Orthopaedics (journal), 159
cockpit voice recorders, 117, *119*; importance of, 4–6, 95–96, 126; noise and, 133–37; popular representations, 129–32, 139–41; surveillance and, 120–22, 125. *See also* ARL Flight Memory Recorder; black boxes; flight-data recorders
Collier's (magazine), 150, 164
The Consequences of Modernity (Giddens), 13
Cooter, Roger, 14, 74
Corliss, George, 78
Cornell Aeronautical Laboratory, 168–71, 188. *See also* Dye, Edward R.
coroners' juries, 55–56, 61
Crash (Beckman), 23, 151, 159, 160
Crash Injury Research Project (Cornell University Medical College), 160–62, 232n68. *See also* De Haven, Hugh
crash-test cinematography, 7–8, 167, 168, 175–87; advantages, 144, 178–85, 186–87, 192–93; data derived from, *192*; difficulties and limitations, 176, 185–86, 193–94; early uses, 176–77, 187; forensic approach and, 23; instrumentation and planning, *181*, *184*, *185*, 185–87; popular representations, 169, *172*, *173*; "second collision" concept and, 189
crash-testing. *See* Crash Injury Research Project (Cornell University Medical College); crash-test cinematography; Institute of Transportation and Traffic Engineering (ITTE, UCLA); Stapp, John Paul
crashworthiness: automobile occupant survivability, 143–44, 159–67, 168, 169–71, 187, 188; data survivability, 115–16, 128. *See also* indestructibility/permanence
Criminal Investigation (Gross), 39, 41, 58–59
criminalistics, 38, 39–40, 41–44, 58, 219n39
cryptography and cryptanalysis, 93

Culture of Accidents (Witmore), 8, 50, 51, 52
Curtis, Scott, 192

dactyloscopy. *See* fingerprinting
d'Alembert, Jean le Rond, 54–55
dangerous instrumentality, 153–54
da Vinci, Leonardo, 107
Death Rode the Rails (Aldrich), 46
deceleration research. *See* Stapp, John Paul
De Haven, Hugh, 159–62, 163, 167, 232n68. *See also* Crash Injury Research Project (Cornell University Medical College)
Derrida, Jacques, 11, 137, 197
Design Paradigms (Petroski), 34, 44
detective fiction, 19, 25, 40–41, 42, 90–93, 205
Detective Fiction and the Rise of Forensic Science (Thomas), 40–41, 42
Dialogues Concerning Two New Sciences (Galileo), 44
Diderot, Denis, 54–55
Doane, Mary Ann, 22–23, 28, 55
Doyle, Arthur Conan, 25, 40, 94, 214
Dupin, Auguste (fictional detective), 40
Duryea, J. Frank, 146
Dye, Edward R., 168, 169–73, 188, 233n98. *See also* Cornell Aeronautical Laboratory
dying words, 112, 137–39, 140–41

Eastman, Joel, 148, 150, 151, 155, 157, 231n52
Eastman Kodak, 94, 224n13
Edison, Thomas: applications of phonograph, 110, 112–14, 137–38, 139; indestructibility/permanence of phonographic record, 107–8, 115, 116; infallibility/unimpeachability of phonographic record, 113; promotional demonstrations, 132–33
The Education of Henry Adams (Adams), 12–13
Eisenhower, Dwight D., 108–9
The Emergence of Cinematic Time (Doane), 22–23
Engineering Failures and Their Lessons (Godfrey), 34
The Engineering Handbook (Dorf, ed.), 38

Engineering Record (journal), 37
Evans-Pritchard, E. E., 9–10, 13
Ewald, François, 13–14, 148–49
experimental physiology, 67–69

faith: in technology, 105, 144, 146–47; in progress, 27–28, 86–87, 147, 218n80; undermining of, 28–30, 60, 86, 147, 218n80. *See also* progress
Faith, Nicholas, 130–31
fallibility, 35–36, 82
Federal Aviation Administration (FAA), 110, 126–27, 133–37. *See also* Civil Aeronautics Board (CAB)
Federal Bureau of Investigation, 140
Feenberg, Andrew, 154, 155
Feld, Jacob, 37
fingerprinting, 41–42
flight-data recorders, 7–8, *96*; automaticity and, 97–101, 104–5, 116, 121, 126; early examples, 97–102, 117, 118, 121, 225n19; importance of, 4–6, 95–96; indestructibility/permanence and, 103–4, 106–7, 109–10, 118, 128, 130, 175, 226n55. *See also* black boxes; cockpit voice recorders; General Mills Ryan Flight Recorder; Vultee Radio Recorder
Flying (magazine), 102, 225n27
Ford Motor Company, 167, 174, 188
forensic desire/impulse, 18, 22, 23, 44, 58, 73, 87, 104, 126–28, 144, 171, 175, 192, 196, 197–98, 210
forensic engineer/engineering (defined), 31–32, 64
Forensic Engineering Investigation (Noon), 31–32
Forensic Engineering: Learning from Failures (ASCE), 31–32
forensic gaze, 39–41, 81–83, 219n29. *See also* autoptic vision; forensics and medical/mechanical pathology
forensic mediation, 6, 21, 24, 30, 98, 190, 199
forensics and medical/mechanical pathology, 16, 37–38, 39, 64. *See also* autoptic vision; forensic gaze

Foucault, Michel, 14, 18, 24, 39
Fox Talbot, William Henry, 209
Fraser, Ken, 126
Freud, Sigmund, 25, 205
Fricke, Lynn: photography's utilities, 202–3, 204–5, 206, 211; photography's infelicities, 206–8, 209
Furnas, J. C., 150–52, 157, 169–71

Gaboriau, Émile, 205, 214
Galileo, 44
General Electric Company, 102
General Mills, 102, 225n36
General Mills Ryan Flight Recorder, *103*; described, 102–3; indestructibility/permanence, 103–4, 106, 108, 109, 226n55; recurrence-prevention goal, 104–5, 106, 128–29, 226n44
General Motors Corporation, 187–88
Giddens, Anthony, 13, 22
Giedion, Sigfried, 27–28, 86
Gilbreth, Frank and Lillian, 189
Ginzburg, Carlo, 25, 212, 213–14
Gitelman, Lisa, 65, 111–12, 215n15
Godfrey, Edward, 34, 37, 59, 60
Goodman, Nan, 53, 54, 59, 221n103
Gordon, James Edward, 37, 62
Gouraud, George Edward, 132–33
graphic method (mechanical-inscription devices), 65–72, *67*, *68*, *69*; automaticity of, 67–68, 71–72, 105; Babbage's self-registering apparatus and, 70–71, 72–73, 81–82, 88; flight-data recorders and, 105, 121, 128; superiority of, 65, 67–68, 69, 71
Green, Judith, 11
Green, Nicholas St. John, 221n103
Gross, Hans, 39–40, 41, 58–60, 221n118
Grossberg, Lawrence, 220n67
Grossi, Dennis, 96
Guthke, Karl, 138–39

Hacking, Ian, 14
Haeusler, Roy, 157

Hall, Stanley, 57–58
Hall, Stuart, 220n67
Hankins, Thomas, 66
Hardingham, Robert, 126
Harper's Weekly (magazine), 146
Hasbrook, Howard, 159
Hayles, Katherine, 153
Haynes, Alex L., 188
Health, Stephen, 210
Hegel, G. W. F., 8, 20
Heidegger, Martin, 1, 193
Helmholtz, Hermann von, 67
Hicks, Joel T., 198, 235n14
high-speed cinematography. *See* crash-test cinematography
history of accidents and failures: post-Industrial Revolution, 12–16, 18–22, 28–30, 44–47, 72–76; pre-Industrial Revolution, 8–12, 48–51. *See also* railroad accidents
history of forensic approach: criminalistics and, 38, 39–40, 41–44, 58, 219n39; forensic medicine and, 16, 37, 39; reason/rationality and, 18–22; technological accidents and, 34–35, 44–47, 52–53, 55–60, 73, 88. *See also* graphic method
History of the Inductive Sciences (Whewell), 196–97
Hobbes, Thomas, 16, 17
Holmes, Sherlock (fictional detective), 25, 40
horse-drawn coaches, 84, *84*–87, *85*, 144, 146–48
Horseless Age (magazine), 146, 147–48
Horwitz, Morton, 54, 59, 221n103
Howard Law Journal, 139
Hume, David, 16–17, 18
Hunter, Louis, 44–45, 46–47
Huskisson, William, 75, 85
Huxley, Thomas, 195–96, 197
Hyman, Anthony, 65–66, 75

Ihde, Don, 33
imaginary projection. *See* prospection
indestructibility/permanence: flight-data recorders and, 103–4, 106–7, 109–10, 118, 128, 130, 175, 226n55; phonograph and, 107–8, 112–13, 115–16; as trope, 106–10
Indestructible Phonographic Record Company, 108
indexicality: authenticity and, 22–23, 81, 129, 132; causal determinism and, 16, 71; detective fiction and, 41; graphic method and, 71, 81; high-speed cinematography and, 182; photography and, 205
Indiana State Police Department, 174
industrialization. *See* faith; history of accidents and failures; history of forensic approach; modernity; progress; railroad accidents
"inoculation" concept, 105–6
Institute of Transportation and Traffic Engineering (ITTE, UCLA), 175–87; collision images, iii, *144, 177, 179, 180, 190, 191, 194*; experimental limitations, 193–94; high-speed cinematography use, 178–86, 189, 192–93; inevitability of accidents, 188, 190; roadway-accident research, 177–78; stunt-driver collision tests, 176–77
instrumental theory of technology, 154, 155
Interstate Commerce Commission, 61
In the Zone (O'Neill), 93–94

Jain, Sarah Lochlann: animal unpredictability, 147; dangerous instrumentality, 153–54; driver-negligence paradigm, 148, 155, 230n15
James, Henry, 27
Job, Macarthur, 126
Johnson, Edward H., 112, 138
Journal of the American Medical Association, 157–58
Journal of the SMPTE, 184–86, 187
Jünger, Ernst, 82–83

Kant, Immanuel, 76, 77–78, 80, 82
Kasson, John, 76, 78–79
Kastenbaum, Robert, 139
Kawin, Bruce, 205

Keenan, Thomas, 32
Kittler, Friedrich: noise, 134, 135; cryptography and cryptanalysis, 93; media storage of remains of people, 89, 141
Koerner, Brendan, 139
Koolhaas, Rem, 15, 16
Kracauer, Siegfried, 13
Krauss, Rosalind, 23
kymograph, 67, *67*

Lacanian Real, 22, 135
Larabee, Ann, 1
Larkin, Brian, 135
last words. *See* dying words
Latour, Bruno, 89–90, 235n30
law and the legal system: dangerous instrumentality and, 153–54; dying words and, 139; forensic-engineering goals and, 31, 32; liability/negligence and, 53–54, 59, 148–49, 181–82; "ordinary care" concept, 221n103; photography and, 204–5
Leviathon (Hobbes), 16
Lévi-Strauss, Claude, 22
Lévy-Bruhl, Lucien, 9, 10, 13
liability: automobile accidents, 148–49, 155–56, 181–82, 191; negligence and, 54, 59, 155–56; technological accidents and, 19, 53–54, 56–57, 59
liberal humanism, 153, 154
lie detectors, 41–42. *See also* polygraphy
Life (magazine), 126, 128
Limpert, Rudolf, 198, *199*
Locard, Edmond, 40
Los Angeles Times (newspaper), 180
Lossier, Henry, 37
Luckin, Bill, 14
Lucretius, 220n83
Ludwig, Carl, 67
Lynch, Michael, 208–9, 235n29

McDermott, Mike, 134
MacGregor, Norval, 90
MacPherson, Malcolm: authenticity of CVR transcripts, 129–30, 136; editing of CVR transcripts, 140, 141; heroization of pilots, 134
Marconi, Guglielmo, 94–95, *95*, 126
Marey, Étienne-Jules, 189; authenticity of graphic method, 69, 72; inventions, *68*, *69*; superiority of graphic method, 65, 67–68, *69*, 71
Marx, Leo, 27, 28, 76, 79
Massachusetts Railroad Commission, 61
Matteucci, Carlo, 67
Mayday (Sturkey), 140–41
Mazlish, Bruce, 27
mechanical-inscription devices. *See* graphic method
mechanist doctrine, 72
mechanization. *See* automaticity; history of accidents and failures; history of forensic approach; modernity; progress
Mechanization Takes Command (Giedion), 27, 86
memory, 6, 118, 202, 205–6
Metaphysics (Aristotle), 49–51
Minifon wire recorder, *122*, *123*, *124*, 134
Mirfield, Titch, *125*, 126
modernity: defined, 6–7; forensic approach and, 6–7, 18–19, 22–23, 25; risk and, 13–14; role of accidents in, 11–12, 18, 44–47, 53; safety discourse, 14–16; technological anxieties, 61–63. *See also* history of accidents and failures; progress; reason and rationality
Modernity and Ambivalence (Bauman), 11
Morelli, Giovanni, 25
Morton, David, 112, 122, 125
Motor Vehicle Accident Reconstruction and Cause Analysis (Limpert), 198
mug shots, 41–42
Mumford, Lewis, 60, 217n72
Murphy's Law, 1–3, 186, 193
Muybridge, Eadweard, 189
myograph, 67, *68*
The Mysterious Black Box (1914 film), 90

Nader, Ralph, 155, 159, 234n134
narrative, 5, 7, 30; causal determinism and,

21–22; reconstruction and, 41, 195; role in accident discourse, 51–52
Nasaw, David, 114
National Safety Council, 149, 155
National Safety News, 103
National Transportation Safety Board (NTSB), 134, 139, 140, 141
Natural History of Religion (Hume), 16–17
Naval Medical Research Institute, 182
negligence, 54, 59–60, 148, 155–56
The New Detectives: Case Studies in Forensic Science (Discovery Channel), 130, 131
Newsweek (magazine), 228n116
New York Times (newspaper): cockpit voice recorders, 126, 135; early flight recorders, 97, 121; flight accidents, 4, 131–32; flight-data recorders, 4, 126, 225n19; phonograph naval applications, 110, 113–14, 115, 120
Nietzsche, Friedrich, 21, 216–17n57
Nisbet, Robert, 26, 217–18n80
noise, 96, 117, 133–37, 140, 141, 208–12
Noon, Randall, 31–32, 44
North American Review (magazine), 107–8, 137–38
Northwestern University Center for Public Safety, 198, 201. *See also* Baker, Kenneth; Fricke, Lynn
Notes on Railroad Accidents (Adams), 56–57, 61
Nye, David, 76

O'Neill, Eugene, 93–94
On the Threshold of Space (1956 film), 164
Oppenheim, E. Phillips, 90–93
Otis, Elisha, 15

Packer, Jeremy, 15
palaetiology, 195–97
paradox of engineering design, the, 33–35, 38, 60
Passages from the Life of a Philosopher (Babbage): broad-gauge rail, 223n27; railway-safety concerns, 74–76, 79; self-registering apparatus, 66, 74, 79, 115, 224n55

The Pathology and Therapeutics of Reinforced Concrete (Lossier), 37
Paz, Octavio, 11, 54
Penley, Constance, 132, 136
permanence. *See* indestructibility/permanence
Peschka, Adolphe, 59
Peters, John Durham, 72, 138
Petroski, Henry: ancient structural failures, 44; human fallibility, 36; paradox of engineering design, 33–34, 35, 38
phonautographs, 66–67
Phonogram (journal), 110–11, *111*, 112–14, 115, 116, 120
phonograph, 110–16, *111*; applications, 110, 112–14, 115, 120, 137–38, 139; indestructibility/permanence, 107–8, 112–13, 115, 116; promotional demonstrations, 132–33
photography: in auto-accident investigation, 160–61, 177–78, 199–200, 202–12; camera as black box, 94; as evidence, 195, 204–5; limitations, 178, 206–12, *207*; Marey and, 68; mug shots, 41–42; photographic excess, 208–12, 235n29
Physics (Aristotle), 49, 51
Piaget, Jean, 9
Picker, John, 132–33
Pingree, Geoffrey, 111–12
Pittsburgh Post-Gazette (newspaper), 140
Plato, 51, 138
Poe, Edgar Allan, vi, 40, 93, 94, 214
polygraphy, 41–42, 68–69, *69*, 88
popular representations of accidents: cockpit voice recordings, 139–41; crash-test research, 164, 165, 169–74, 180–81; early automobile accidents, 150–52, 169–71; early technological accidents, 46, 73–74
popular representations of forensic evidence, 32–33, 129–32, 139–41. *See also* MacPherson, Malcolm
Popular Science (magazine), 180–81
Porter, Dennis, 41, 43
positivism, 58, 68
preservation, 108–9. *See also* indestructibility/permanence

prevention. *See* recurrence prevention
progress, 26–30; defined, 26–27, 217n72; faith in, 27–28, 86–87, 147, 218n80; forensic approach as contributing to, 8, 29–30, 33–35, 38, 60–61, 82–83, 114; technological sublime and, 79
prospection, 6, 24–25, 30, 43, 81–82, 87–88, 114–15
Public Safety (journal), 160–61

Quick, Charles, 139

Rabaté, Jean-Michel, 212
Radio News, 99–101
railroad accidents, 35, 57, 73, 75; vs. horse-drawn coaches, 85–87, 146, 147; investigation of, 56–58; liability and, 19, 53, 56–57; public anxiety, 45–46, 73–74. *See also* Babbage, Charles; boiler explosions
Railroad Accidents: Their Causes and the Means of Preventing Them (With), 19
Railroad Gazette, 56
railroads, technological sublime and, 78–79
Railway Detectives (Hall), 57–58
Railway Inspectorate, Her Majesty's, 56–58, 61, 87
Ray, Robert, 209, 211
Reader's Digest (magazine), 150–52, 157, 169–71
reason and rationality: accidentality and, 10–11, 12, 13–14, 17–18, 50–51, 82, 153, 185–86, 194; audile technique and, 135; causal determinism and, 10–11, 14, 17–18, 19–21, 197; detective fiction and, 25, 41; forensic approach and, 22, 29, 38, 73, 81, 87–88, 197–98; human subjectivity and, 153, 154; instrumentalist rationality, 154–55; vs. photographic excess, 210; progress and, 82–83, 147; technological sublime and, 77–78, 80, 82, 87. *See also* progress
reckless drivers. *See* accident discourse (human-error focus)
reconstruction (generally), 7–8, 195; causal determinism and, 59, 97, 98; of crimes, 40–44, 58; defined, 32, 42; forensic temporality and, 24; historical background, 58
reconstruction of auto accidents, 197–212; investigation process, 198–206, 210–11; photography limitations, 206–12, *207*; photography uses, *199*, *200*, 202–6, *203*, *204*; scene sketches and diagrams, 200–202, *201*, *202*
recurrence prevention, 97–98, 102; causal determinism and, 7, 59–60, 61, 80, 96, 105, 118, 128–29; forensic temporality and, 43, 87; as goal of flight-data recorders, 96, 100, 104–5, 106, 128–29, 226n44; as goal of forensic approach, 57, 59–60, 118; as goal of maritime phonograph, 111, 114
redundancy, 186, 188, 193
Reed, Robert, 46, 73–74
reminiscence, 42–43, 55
repetition. *See* time/temporality
representation, 7, 22–23, 43, 206–7. *See also* graphic method; photography
reproduction, 24; photographic, 202, 209; sonic, 112–15, 121, 126–27
retrospection, 23, 42–43; prospection and, 6, 24–25, 81–82, 114–15; retrospective prophecy, 195–97
risk, 1, 6, 13–14, 15, 36, 46, 52–53, 62, 114, 129, 151–52, 185, 198
Risk and Misfortune (Green), 11
Rivers, R. W., 198–99, 200, 203, 204, 207
Ronell, Avital, 136–37, 143
Rosenheim, Shawn James, 93
Ryan, James J., *103*; auto-safety research, 174–75; flight-recorder description, 102–3; flight-recorder indestructibility/permanence, 103–4, 108, 109; recurrence-prevention goal, 104–5, 128–29, 226n44. *See also* General Mills Ryan Flight Recorder

safety discourse, 8, 14–16; Babbage's concerns, 74–75, 83–84, 87; boiler explosions and, 47. *See also* accident discourse; Institute of Transportation and Traffic Engineering (ITTE, UCLA); recurrence prevention; Stapp, John Paul

258 INDEX

Sawday, Jonathan, 39
Scarry, Elaine, 195
Schaffer, Simon, 23
Schivelbusch, Wolfgang: pre-industrial vs. technological accidents, 55; shock caused by technological accidents, 28, 52–53, 86, 221n97; technological anxiety, 45–46, 74; technological security, 146
Schnapp, Jeffrey, 84–85
Science Digest (magazine), 102
Science (magazine), 97–98, 121, 225n19
Scientific American (magazine), 138
Scott de Martinville, Édouard-Léon, 66–67, 72
Sear, Lane, 126
second collision, 189–90, 234n134
security, sense of, 52–53, 77, 86
seismography, 88
Sekula, Allan, 219n39
self-registering apparatus (Babbage), 7–8; adoption of, 224n55; data survivability and, 115; described, 65–66; future perfectibility and, 87–88; goals of, 66, 74, 83–84; graphic method and, 70–71, 72–73, 81–82, 88; technological sublime and, 79, 80–81, 87
September 11, 2001, terrorist attacks, 134, 140, 228n116
serendipity, 211, 212–14
Serres, Michel, 89–90, 133, 135, 136, 229n127
Severy, Derwyn M.: goal of ITTE crash-tests, 175–76; high-speed cinematography, 182–86, 187, 189; inevitability of accidents, 188–89, 190; roadway-accident research, 177–78; stunt-driver crash tests, 176–77
Shannon, Claude, 133
Shapin, Steven, 23
Shifting the Blame (Goodman), 53, 54
Signal and Noise (Larkin), 135
Silverman, Robert, 66
Simmel, Georg, 13
Singer, Ben, 13, 46
Sneddon, James, 201
Socrates, 138
Sontag, Susan, 204, 211, 236n41
speculation, 24–25, 43, 196–98
sphygmograph, 67, 68

Spivak, Gayatri Chakravorty, 229n129
Stapp, John Paul: automotive crashworthiness, 164–66, 167, 188; celebrity, 165, 171; Edwards Air Force Base experiments, 1–2, 2, 3, 6, 23, 162–63, 163, 182; Holloman Air Force Base experiments, 163–64
statistical probabilism, 11, 14
steamboat accidents, 44–45, 45, 46–47, 52–53, 55–56
Steamboats on the Western Rivers (Hunter), 44–45, 46–47
Stephenson, George, 34
Stephenson, Robert, 34–35, 60
Sterne, Jonathan, 107, 108, 135
Straith, Claire L., 156–57, 158, 160
Structures: Or Why Things Don't Fall Down (Gordon), 37
Studebaker Corporation, 157
Sturkey, Marion, 140–41
Styling vs. Safety (Eastman), 148, 150, 151, 155, 157
Success through Failure (Petroski), 34
superstition, 9–11, 13, 22
surveillance, 120–25, 126
survivability. *See* crashworthiness; indestructibility/permanence

Tainter, Charles Sumner, 113
technological sublime, 76–79, 80–81, 82, 87
Technology and Culture (journal), 38
telegraphy, 92–93, 122
Ten Books on Architecture (Vitruvius), 44
"The Three Princes of Serendip" (Tramezzino), 212–14, 213
Thomas, Ronald, 40–41, 42
Thorwald, Jürgen, 39–40, 41, 58
Tichi, Cecelia, 63
Time (magazine), 163, 164, 165
time/temporality, 22–24, 42–44. *See also* prospection; retrospection
Titanic, the, 28, 63, 107
To Engineer is Human (Petroski), 33, 36
To Forgive Design (Petroski), 34
traces. *See* graphic method
Trachtenberg, Alan, 86

Traffic Accident Investigators' Manual (Rivers), 198–99, 200, 203, 204, 207
Traffic Collision Investigation (Northwestern University Center for Public Safety), 201, 202–3, 204–5, 206–8, 209, 211
Tredgold, Thomas, 31, 218n21

UCLA (University of California, Los Angeles). *See* Institute of Transportation and Traffic Engineering
United Data Control, 126–27

Valverde, Mariana, 219n29
Van Kirk, Donald, 199, 201, 202, 205, 206
van Wyck, Peter, 220n75
Vehicular Accident Investigation and Reconstruction (Van Kirk), 199, 201, 202, 205, 206
Vierordt, Karl, 67
Virilio, Paul, 29, 47–49, 52, 220n75, 221n97
Vitruvius, 44
Voltaire, 195, 197, 213–14
Vultee Radio Recorder, 98–101, *99*, *100*, 126

Wallace, DeWitt, 151–52
Walpole, Horace, 214
Ward, Joseph, 31, 37–38, 44
War Medicine (journal), 159–60
Warren, David. *See* ARL Flight Memory Recorder

Watt, James, 69
Wayne State University, 174
Weber, Wilhelm, 66
Weizman, Eyal, 32
Wertheim, Guillaume, 66
Wheatstone, Charles, 66
Whewell, William, 196–97
Wiener, Norbert, 89, 133
Williams, Raymond, 223n42
Winner, Langdon, 154–55
witchcraft, 9–10
With, Emile, 19
Witmore, Michael, 8, 50, 51, 52
Wittgenstein, Ludwig, 222n14
Woman's Day (magazine), 169–71, 172–74, 233n98
Woodward, Fletcher D., 157–58, 161–62, 231n54
World's Fair (1853), 15
World War I, 27, 63
World War II, 27–28
Wren, Christopher, 69
Wright, Patrick, 226n48

Young, Thomas, 66

Zadig (Voltaire), 195, 197, 213–14
Žižek, Slavoj, 22, 28